JX$16·50

Fundamentals of
Data Communications

Fundamentals of Data Communications

Jerry FitzGerald

Tom S. Eason

JOHN WILEY & SONS
New York • Chichester • Brisbane • Toronto

A Wiley/Hamilton Publication

This book was set in 10 point
Aster by Beacon Typesetting. Mike
Rogondino designed the text and
cover, Graphics Two drew the art
and Susan Welling was the copyeditor.
Printing and binding was by Halliday
Lithograph and Chuck Pendergast
supervised production.

Library of Congress Cataloging in Publication Data

FitzGerald, Jerry   1936–
      Fundamentals of data communications.

      "A Wiley/Hamilton publication."
      Bibliography: p.
      Includes index.
      1. Data transmission systems. I. Eason, Tom S.,
joint author. II. Title.
TK5105.F58   621.38'041   77-20842
ISBN 0-471-26254-4

Printed in the United States of America
10 9 8 7 6 5 4 3

About the Authors

Dr. Jerry FitzGerald is the founder of Jerry FitzGerald & Associates, a management consulting firm located in Redwood City, California. He has had extensive experience in data communications, data processing security, and EDP auditing.

Dr. FitzGerald has been active in the review and validation of distributed versus centralized on-line systems and has worked on numerous data communications and telecommunications network problems. He has taught data communications training courses and seminars at the University level and to practitioners in the field. He is also a nationally known expert in EDP security and EDP auditing and control. Prior to establishing his own firm he was a Senior Management Consultant with SRI International (formerly Stanford Research Institute).

Before joining SRI he was an Associate Professor in the California State University and Colleges System teaching courses in data communications, data processing, and EDP auditing at Cal State Poly University at Pomona, Cal State University at Hayward, and the University of California at Berkeley. Dr. FitzGerald's educational background includes a PhD in Business Administration, a Masters degree in Economics, an MBA, and a Bachelors degree in Industrial Engineering. He is the author of a textbook on systems analysis as well as numerous articles.

Tom S. Eason is Assistant Director of the Information Systems Management Department of SRI International in Menlo Park, California. He has a total of 26 years experience in data processing and communications. Prior to his association with SRI, he held management and senior technical positions at Planning Research Corporation and in the aerospace industry. Among his technical and managerial achievements in the field of data communications are the design and implementation of the first international hotel reservation system, pioneering work in the design of problem-oriented languages, and consultation for many of the leading industrial users of data communications in the U.S. and abroad.

This book draws on his experience in both government and industrial communications-oriented data processing systems, and sets forth design approaches which have been used with considerable success in these environments.

Preface

The primary objective of this book is to support the teaching of a first course in data communications at the college or university level. Since the techniques and ideas have been tested successfully in actual practice by the authors, the book can also serve as a valuable reference to the practical aspects of developing and designing data communication networks. The book is based on a data communications course that Dr. FitzGerald teaches at the university level and to corporate and government data processing personnel who want to learn how a data communication network operates.

A unique feature of the book is Chapter 8: Designing Communication Networks. After reading the necessary technical and informational preparation presented in Chapters 1 to 7, the beginning practitioner will find in this chapter a step-by-step approach to designing a data communication system. There are sections on identifying and defining the problem, designing the proposed system, analyzing the types of messages, determining the total traffic, developing alternative configurations, calculating the network costs, as well as on the implementation and follow-up evaluation. This chapter, along with the network design problems, will equip the student with the necessary knowledge to design basic data communication networks.

The preceding chapters, 1 to 7, build the foundations for the design chapter. They include a chapter on fundamental communication concepts, such as synchronous/asynchronous transmission, modulation, and half and full duplex; a chapter that discusses and defines the hardware of data communications; as well as a chapter on communication techniques such as multiplexing, multipoint line control, and switching. Other chapters discuss programming data communication systems, error detection and correction in data communications, and tariffs and costs of data communication lines. To guide the reader in evaluation and to increase the realism of the analyses, costs for equipment and services are included at the levels prevailing in late 1977. The designer should verify costs when conducting "real-life" analyses.

The book is at once practical, general, and specific. It is ideally suited for use with outside materials such as current periodical articles, individual or class case problems, or real-life network problems. In our experience, we have found that data communications design training material can be meaningful only if the student is required to work and think his way through it. Thus, through the organization of the chapters, we have attempted to structure the text in a real-world style that builds up to the chapter on designing data communication networks. In this way, we hope to broaden the student's appreciation for independent curiosity, exploration, and creativity, and especially we hope to make the student a good data communications analyst.

Redwood City, California

Jerry FitzGerald
Tom S. Eason

Contents

Fundamentals of
Data Communications

1

Introduction to Data Communications

This chapter serves two purposes: it introduces the book and it introduces the subject. In the first, the purpose, scope and structure of the book are described. In the second, the term "data communications" is defined, and its major applications are enumerated and described.

Purpose and Scope of This Book

Data communications is a complex subject, but this is not a complex book. Most books in this field are complex, and rightly so, because they are intended as reference sources for the experienced systems designer. This book requires no prior experience in data communications, voice communications, or in electronic engineering. Rather, this book assumes a basic grounding in data processing and a desire to complement this background with a general knowledge of data communications. After completing a course of study based on this book, you should be able to:

- Understand the alternatives available to you in hardware, software, and transmission facilities.
- Put that understanding to work by making informed decisions among·these alternatives.
- Integrate these decisions into a cohesive data communication system design, and carry it forward into reality.

• Perform the above activities for systems of increasingly greater scope and complexity as you build experience, judgment, and confidence.

How This Book Is Structured

This book is divided into eight chapters, of which this is the first. Chapter 2 discusses fundamental communications concepts, such as how data are transmitted over telephone lines, types of transmission, and techniques of modulation. Chapter 3 gives the reader a complete and thorough picture of data communication hardware components utilized in creating today's data communication networks. Chapter 4 explains the techniques that are utilized to control transmission of data between sending and receiving stations. Chapter 5, Error Detection and Correction, discusses the various sources of errors, and effective methodologies for the detection and correction of errors in the data communication system. Chapter 6 discusses the role that software plays in data communication systems, depicts examples of communication software, and presents software design and testing guidelines for data communication programming.

After completing Chapters 1 to 6 the reader will have a basic understanding of fundamental data communication network concepts, hardware, control techniques, error detection/correction, software characteristics, and can then begin to apply this knowledge to the design of data communication networks. Chapters 7 and 8 are the basic design chapters. Tariffs and Common Carriers (Chapter 7) describes the types of communication facilities that are available to the designer and the costs (tariffs) of each of these services. Designing Communication Networks (Chapter 8) addresses the primary goal of this book. With the aid of the principles and examples given in this chapter, the systems designer should be able to develop basic designs and cost/benefit analyses of data communication systems.

Chapter 8 describes and illustrates the process of a system design. In particular, it leads the reader through the 10 basic steps of system design:

Step 1:   Define the problem so it is evident and understandable to management, the potential users, and the system design personnel. All these parties must work toward a solution to the same problem. The solution should be stated in a positive manner as a course of action that management can take.

Step 2:   Prepare an outline of the approach and methodology that you are going to utilize in studying and designing the proposed system, in order to organize a detailed plan of action. The outline should be based on Steps 3 through 8 of this sequence, adjusted to the particular problem in hand.

Step 3:   Gather general background information on the areas that will be affected by the proposed data communication system. It is imperative that the system designer know the background of the industry, company, or government agency, and the various individual areas that will be affected because every aspect of the business is integrated to work together to accomplish the organization's objectives.

Step 4:   Study the interactions between the areas that are affected by the proposed data communication system. Learn the interactions between the various departments or agencies within the organization. These interactions should be defined in terms of the outputs and inputs of each organizational unit and the processes performed by the various organizational units, insofar as they affect, or are affected by, the design.

Step 5:   Obtain a general understanding of the existing system. The designer must understand the existing sytem, whether it is a manual system or an on-line data communication system that is being replaced or enhanced. It is important in the design process for the designer to know why certain things are done in the way they are.

Step 6:   Define the proposed system's requirements, in order to assemble an overall picture of the system. These requirements must be defined within the framework of the goals and objectives of the entire organization as well as of each department or agency. Strive to make these requirements quantitative and detailed.

Step 7:   Using the requirements defined in Step 6, design the proposed data communication system. If the breadth of the requirements allow it, design various alternative systems, taking different technical and/or procedural approaches. Consider the maintenance and network management aspects of each alternative system.

Step 8: Develop cost comparisons for the alternative designs. Select the network configuration alternative that best meets both the cost and system requirements. Make minor adjustments to optimize the design, if possible.

Step 9: Sell the system. In this important step, the system designer must convince both management and the users that it is cost-effective to implement the data communication system design and that it will produce the desired results.

Step 10 Implementation, follow-up, and reevaluation are mandatory steps if the system is to be widely accepted by its users. In the implementation process, the system designer participates in the procurement, development, and installation of the system, as well as in organizing the maintenance procedures and training its users. During follow-up, the system designer observes the actual operation and makes sure that all parts of the new system are operating according to specifications. Reevaluation takes place after the system has been broken in; the systems designer returns to evaluate its efficiency and make whatever changes are needed to optimize its performance.

In summary, this book is structured to expose the reader to each basic area of knowledge in data communications and to integrate the areas by ending with an extensive chapter on designing a communication network.

Other Sources of Information

There are many good books and periodicals on data communications. The bibliography at the end of this book lists a representative sample. Also at the end of this book is a glossary of data communication terms. Use it regularly and carefully to enhance your understanding of data communications. At this point review the bibliography and the glossary.

Definition of Data Communications

Data communications is the movement of encoded information from one point to another by means of electrical* transmission systems. Such systems are often called data communication net-

---

*In a few years we will probably have to substitute for this word the phrase "electrical, optical, and electro-optical."

works. In general, data communication networks are established to collect data from remote points (called terminals) and transmit that data to a central point equipped with a computer or another terminal, or to perform the reverse process, or some combination of the two. Data communication networks facilitate more efficient use of central computers. They improve the day-to-day controls of a business by providing faster information flow. They provide message switching services to allow terminals to talk to one another. In general, they offer better and more timely interchange of data among their users and bring the power of computers closer to more users. The objectives of most data communication networks are to:

- Reduce the time and effort required to perform various business tasks
- Capture business data at its source
- Centralize control over business data
- Effect rapid dissemination of information
- Reduce current and future costs of doing business
- Support expansion of business capacity at reasonable incremental cost as the organization grows
- Support organizational objectives in centralizing or decentralizing computer systems
- Support improved management control of the organization.

Uses of Data Communications

While data communications might be used in many different situations, business operations that exhibit some of the following characteristics usually can benefit from the use of a data communication network:

- Decentralized operations
- A high volume of organizational mail, messenger service, or telephone calls between the various organizational locations (the voice communication corridors, i.e., telephone calls, may become or be replaced by the data transfer corridors).
- Repetitive paperwork operations, such as the re-creation or copying of information.
- Inefficient and time-consuming retrieval of current business information

- Slow or untimely handling of the organization's business functions
- Inadequate control of the organization's assets
- Inadequate planning and forecasting.

Types of data communication systems that can correct the above deficiencies fall into seven categories:

- Source data entry and collection
- Remote job entry (RJE)

Data Communications Usage Modes	Examples of Applications	Typical Characteristics of Transactions
Source data entry and collection	Sales status data; Inventory control; Payroll data gathering;	Transactions collected several times per day or week, direct response message not issued for every transaction
	Point-of-sale system; Airline reservations	Transactions arrive frequently (every few seconds) and demand response within a few seconds
Remote job entry (RJE)	Remote high-speed card reading and printing; Local access to distant computer power	Transactions usually are bunched and require processing times ranging from minutes to hours. Input and output for each transaction may take seconds or minutes.
Information retrieval	Credit checking; Bank account status; Law enforcement; Government social services; Hospital information systems; Bibliographic systems	Relatively low-character volume per input transaction, response required within seconds. Output message lengths usually short but might vary widely with some types of applications.
Conversational time sharing	General problem solving; Engineering design; Calculations; Text editing	Conversational response required, within a few seconds
Message switching	Company mail delivery and memo distribution	Delivery time requirements range from minutes to hours
Real time data acquisition and process control	Numeric control of machine tools; Remote meter and gauge reading	Remote sensors are continuously sampled and monitored at widely varying time intervals.
Interprocessor data exchange	Processor, program, and file sharing applications of all types involving communications between computers	Infrequent, burst arrivals consisting of large data blocks requiring transmission to another CPU, usually within milliseconds.

Figure 1-1:   Categories of Data Communications

- Information retrieval
- Conversational time sharing
- Message switching
- Real time data acquisition and process control
- Interprocessor data exchange

Figure 1-1 summarizes many important characteristics of these typical uses of data communications, giving specific application examples and typical transactions for each application. This tabulation is the framework within which we will work for the remainder of this book. Study this figure carefully. Using your knowledge of the functional characteristics of the specific application examples, consider the information given in the "typical characteristics of transactions" column. Observe how these characteristics change from one usage to the next.

Basic Components of a Data Communication System

The three basic components of a data communication system are the source, the medium, and the sink. The source is the originator of the information. The medium is the path through which the information flows. The sink is the mechanism that accepts the information. In this definition, a terminal often alternates as both a source and a sink. The medium is nothing more than the communication lines ("circuits") over which the information travels. In all the cases with which we are concerned, the lines are leased from a "common carrier" such as the Bell Telephone System. A common carrier is a company recognized by the Federal Communications Commission (FCC) or an appropriate state licensing agency as having the right to furnish communication services to individual subscribers or business organizations.

Figure 1-2 depicts the data communication process. Five elements are included: Two terminals, which function as sources and sinks, the transmission facility, and two devices called modems, which form the interfaces between the terminals and the transmission facility. A terminal is, in fact, an input/output device such as a teleprinter, a video (CRT)* device, a RJE unit, or the like. The medium shown in Figure 1-2 is the transmission facility which, in reality, is usually the telephone line. The modem in Figure 1-2 (explained in more detail in Chapter 3 and the glossary) is a device that converts data from a terminal to a form of data that can be transmitted over telephone lines and vice versa.

---

*Cathode Ray Tube

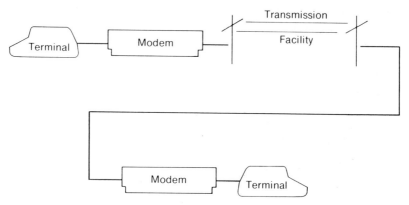

Figure 1-2:   Data Communication Process

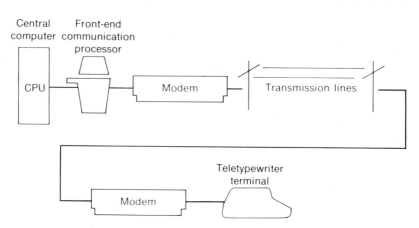

Figure 1-3:   Data Communication Network

Figure 1-3 depicts one type of point-to-point data communication network. A point-to-point line connects one terminal or computer to one other terminal or computer. In this figure, there is a central computer with a front-end communication processor. Again, there are modems at each end of the transmission line to convert the data from a form that is understood by computers and terminals to a form that can be sent over transmission lines. There is also a teletypewriter terminal that is utilized for input and output of the data. Figure 1-4 depicts a data communication network that has a video (CRT) terminal. The only difference between Figures 1-3 and 1-4 is in the type of remote terminal. The basic components of each data communication system are the same. More information on these pieces of hardware will be presented in Chapter 3, Data Communication Hardware.

Also, study the other network configurations shown in Figure 1-5. Of the seven configurations shown, 1, 2, and 7 are point-to-point. Configuration 7 depicts the use of dial-up lines as a backup should the leased transmission lines fail. Configurations 3 and 4 depict multidrop lines, with 3 using leased lines and 4 using a privately owned cable. Configuration 5 depicts a multiplexed setup, and 6 depicts the use of a remote minicomputer to control a group of terminals.

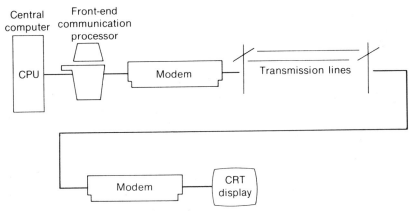

Figure 1-4:   Network with a Cathode Ray Tube (CRT) Display

Questions—Chapter 1

This first chapter is designed to give some idea of the scope of data communications and the practical purposes to which data communications can be put. In the world of business, industry, and government, the systems analyst who is aware of ways that data communications can be put to work will find a variety of opportunities to use his awareness and skills. Applications of this sort are the *most rapidly expanding field* in electronic data processing. The questions that follow are designed to sharpen (and test) your awareness of applications opportunities. Accept the challenge of responding to these questions thoughtfully, insight-

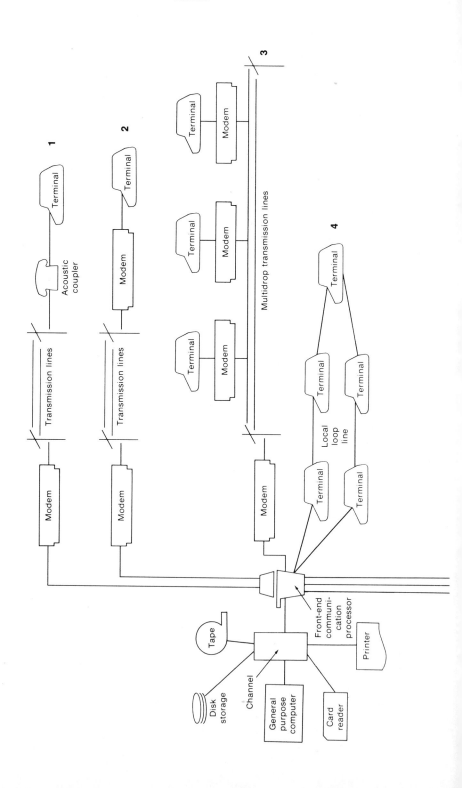

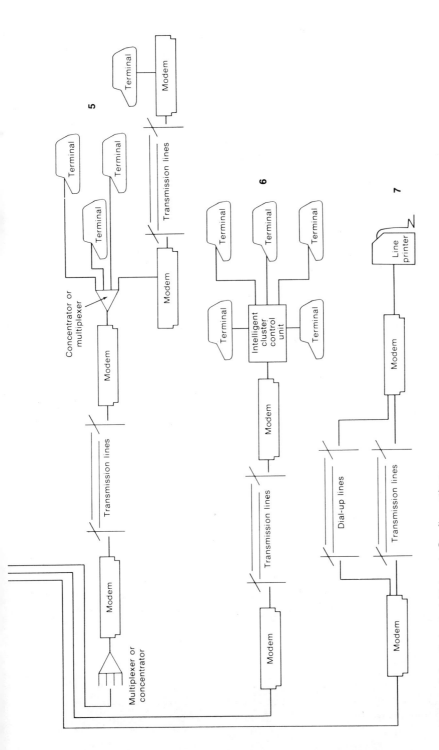

Figure 1-5: Selected Network Configurations

fully, and thoroughly, recognizing that you are laying the foundation for your knowledge in this field.

True or False

1.   A transmission facility is a device that converts data from a terminal to a different form.

2.   The same device can function as both a source and a sink in some data communication systems.

3.   Data communication networks are only established to accomplish the collection of data from remote points.

4.   One of the basic steps in data communication systems design is to convince both management and the users that it is cost-effective to implement the data communication system and that it will produce the desired results.

5.   Most data communication networks share the objective of increasing the time and effort required for performing various business tasks.

6.   It is likely that businesses with decentralized operations can benefit from the use of a data communication network.

Fill-in

1.   In a data communication network, the ——— is the path through which information flows.

2.   In a data communication network, the ——— is the originator of the information.

3.   In a data communication network, the ——— is the mechanism that accepts the information.

4.   A ——— ——— is a company recognized by the Federal Communications Commission or appropriate state licensing agencies as having the right to furnish communication services to individual subscribers or business organizations.

5.   RJE stands for ——— ——— ———.

6.   CRT stands for ——— ——— ———.

7.   A CRT is a type of———.

8.   Data communication is the movement of ——— ——— from one point to another by means of electrical transmission systems.

9.   —— basic steps of system design are enumerated in the text.

Multiple Choice

1.   Credit checking is an example of an application illustrating which of the following modes of data communication usage?
   a) Conversational timesharing
   b) Information retrieval
   c) Real time data acquisition and process control
   d) Source data entry and collection
   e) None of the above

2.   A point-of-sale system is an example of an application illustrating which of the following modes of data communication usage?
   a) Conversational timesharing
   b) Information retrieval
   c) Real time data acquisition and process control
   d) Source data entry and collection
   e) None of the above

3.   Company mail delivery and memo distribution is an example of an application illustrating which of the following modes of data communication usage?
   a) Conversational timesharing
   b) Information retrieval
   c) Real time data acquisition and process control
   d) Source data entry and collection
   e) None of the above

4.   Numerical control machine tools is an example of an application illustrating which of the following modes of data communication usage?
   a) Conversational timesharing
   b) Information retrieval
   c) Real time data acquisition and process control
   d) Source data entry and collection
   e) None of the above

5.   Engineering design calculations and text editing are applications illustrating which of the following modes of data communication usage?
   a) Conversational timesharing
   b) Information retrieval

c) Real time data acquisition and process control

d) Source data entry and collection

e) None of the above

6.   Which of the following characteristics is typical of transactions associated with information retrieval?

a) Delivery time requirements range from minutes to hours

b) Infrequent, burst arrivals consisting of large data blocks requiring transmission to another CPU

c) Transactions collected several times per day or week, but a direct response message is not issued for every transaction

d) Transactions usually bunched and require processing times ranging from minutes to hours

e) None of the above

7.   Which of the following characteristics is typical of transactions associated with message switching?

a) Delivery time requirements range from minutes to hours

b) Infrequent, burst arrivals consisting of large data blocks requiring transmission to another CPU

c) Transactions collected several times per day or week, but a direct response message is not issued for every transaction

d) Transactions usually bunched and require processing times ranging from minutes to hours

e) None of the above

8.   Which of the following characteristics is typical of transactions associated with interprocessor data exchange?

a) Delivery time requirements range from minutes to hours

b) Infrequent, burst arrivals consisting of large data blocks requiring transmission to another CPU

c) Transactions collected several times per day or week, but a direct response message is not issued for every transaction

d) Transactions usually bunched and require processing times ranging from minutes to hours

e) None of the above

9.   Which of the following characteristics is typical of transactions associated with remote job entry?

a) Delivery time requirements range from minutes to hours

b) Infrequent, burst arrivals consisting of large data blocks requiring transmission to another CPU

c) Transactions collected several times per day or week, but a direct response message is not issued for every transaction

d) Transactions usually bunched and require processing times ranging from minutes to hours

e) None of the above

Short Answer

1. In the real world there are many examples of the source/medium/sink relationship shown in Figure 1-2. Take at least one entry from the "Examples of Applications" column of Figure 1-1 for each of the first three usage modes, and identify some of the source/medium/sink relationships that apply.

2. In the section "Uses of Data Communications," there is a list of the characteristics of organizations that can benefit from data communication systems. Examine these characteristics and identify who (or what function) in the organization will benefit and describe how.

3. In the section "Definition of Data Communications," there is a list of eight objectives common to most data communication networks. With reference to these, of what value is it to capture business data at its source? How can a data communication network centralize control over business data?

4. Using Figure 1-1 as a source of ideas, find a terminal of a data communication network convenient to you. Observe the operator in action, interview the operator, find out what the terminal communicates with, how it communicates, and over what medium it communicates. Write a brief description of the purpose, operation, and characteristics of the data communication system that you have observed.

5. After completing a course of study based on this textbook, what should the student be able to do?

2

Fundamental Communications Concepts

The preceding chapter introduced the study of data
communications by describing a few basic ideas and
showing the general areas in which such systems
are applied. This chapter begins the study of data communications
technology by concentrating on what happens when data are moved
from point to point. This requires discussion
of some topics that verge on electrical engineering;
however, these discussions are phrased in
laymen's terms. It is not necessary to have
an engineering-level understanding of the topics to be
an effective analyst of data communication
applications. It is important, however, to understand
these concepts and to make intelligent use of them. The
content of this chapter is basic; knowledge of the
content is prerequisite to your competence in this field.

Modes of Transmission

Systems that transmit data must have consistent methods of
transmission over communication channels. All the systems that
we will discuss transmit binary data, or data forms that are
intrinsically binary. Basically, binary data can be sent over communication lines in either serial or parallel mode. The internal

transfer of data within modern computers is done in parallel mode. In other words, if the internal structure of the computer uses an 8-bit code element, then all 8 bits of an element are transferred between the main memory and any operational register in the same computer cycle.

This type of transfer is not normally used in data communications. In data communications, transfer of information is usually done in serial mode. Serial transmission implies that a stream of data is sent over a communication line bit by bit. For example, the bits of information in an 8-bit USASCII* code may be sent down a transmission line (channel) as shown in Figure 2-1.† The USASCII code is used as an example in the following figures; it will be discussed further in the next section on coding terminology and structures.

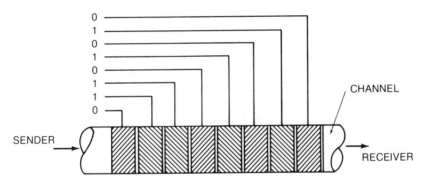

Figure 2-1:   Serial Transmission of an 8-Bit USASCII Code

Parallel transmission, although seldom used outside the computer, is illustrated in Figure 2-2. This figure shows how all 8 bits of the USASCII code travel down a channel simultaneously, followed a short time later by eight more bits. The distinguishing difference between serial and parallel transmission is that, in serial transmission, the transmitting device sends a bit followed

*United States of America Standard Code for Information Interchange.

†Figure 2-1, and several following figures, are used to depict forms of transmission. These are conceptual pictures only—they are not representative of the electrical engineering facts of data transmission. Use them to help your understanding of the concepts only.

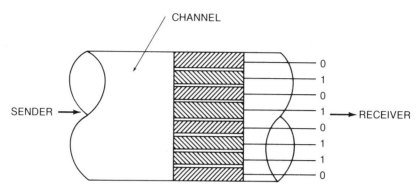

Figure 2-2: Parallel Transmission of an 8-Bit USASCII
            Code

by a time interval, then a second bit, and so on until all the bits
are transmitted. It takes n time cycles to transmit n bits (8 in the
case of USASCII). In parallel transmission, n bits are sent, followed by a time interval, then n bits are sent, and so on. In
parallel transmission, then, n bits are sent in one time cycle,
whereas in serial transmission the same n bits take n time cycles.

   Parallel transmission is a method of transfer in which all the
bits of a character are sent simultaneously either over separate
lines or on different frequencies of the same line (the section on
multiplexing in Chapter 3 will explain how this is done). Parallel
transmission uses a low-cost transmitter, but the receiver is a
high-cost item, and it generally also requires higher cost transmission lines than does serial transmission. It is not used on
low-speed lines because its primary purpose is to speed up the
transmission between two points. Neither is it used on long-
distance lines because the bits drift back and forth in time
relation to one another and may interfere with the bits of the
preceding or following character. With parallel transmission, the
line speed, in bits, is really the line speed in characters since all
the bits of a character are sent simultaneously.

   Most data communications are performed by serial transmission. Three transmission modes are in common usage: asynchronous, synchronous, and isochronous transmission. Note that
all three of these modes of transmission are serial. Figure 2-3
illustrates asynchronous transmission.

   The first mode of transmission, asynchronous, is often referred to as start-stop transmission. This is because the transmitting device can transmit a character at any time that it is

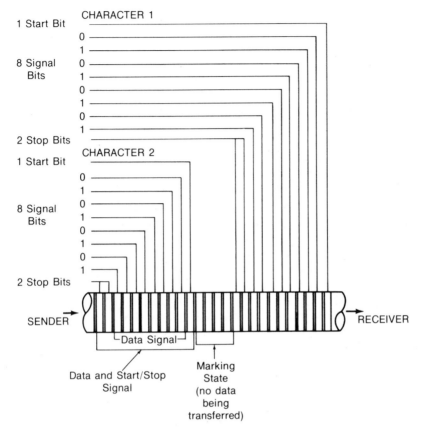

Figure 2-3: Asynchronous Transmission of an 8-Bit
USASCII Code in Serial Mode

convenient and the receiving device will accept that character. Characters can be sent at irregular intervals: for example, 1 character per second or 1 character and then a 10-second wait. To enable the receiver to recognize a character when it arrives, each of the characters that are transmitted have a start bit preceding, and one or two stop bits following the data signal bits. In Figure 2-3, the data signal bits are the 8 bits of the USASCII code, and as can be seen, this code has one start bit and two stop bits (sometimes USASCII is transmitted using only one stop bit). In other words, in asynchronous transmission each character is synchronized by its own start and stop bits. Some coding structures have one start bit and one stop bit, some have one start bit and two stop bits, and a common form of Baudot code has one start bit and one "long" stop bit equivalent to 1.42 normal bits. A complete discussion of coding structures is provided in the next section, Coding Terminology and Structure.

In summary, a start bit is a signal that is used to inform the receiving terminal to start sampling the incoming data signal at a fixed rate so it can be interpreted into its proper character structure. A stop bit, which follows the data signal bits, informs the receiving terminal that a character has been received and resets the terminal for recognition of the next start bit. Synchronization of the terminals is reestablished upon the reception of each character.

The second transmission mode, synchronous, is used for the high-speed transmission of a block of characters. In this method of transmission, both the sending device and the receiving device are operated simultaneously and are resynchronized after each few thousand data signal bits are transmitted. Start/stop bits are not required for each character. Synchronization is established and maintained either when the line is idle (no data signals being transmitted) or just prior to the transmission of a data signal. This synchronization is established by passing a predetermined group of "sync" characters between the sending and the receiving devices. Figure 2-4 shows how the data signals are contiguous and how one long stream of data bits is transmitted from the sending to the receiving device. In other words, the sending device will send some "sync" characters to the receiving device so the receiving device can determine the time frame between each of the bits.

The sending device sends a long stream of data bits that may have thousands of bits. The receiving device, knowing what code is being used, counts off the appropriate number of bits and assumes this is the first character and passes it to the computer. It then counts off the second character and so on. If the code used is USASCII, the receiving device counts off the first 8 bits and sends them to the computer as a character; it then counts off the second 8 bits and sends them as a character, and so on.

Synchronous transmission is more efficient in that there are fewer control bits in proportion to the total number of bits transmitted. The synchronization may take only 16 to 32 bits, while the stream of bits for the data signal may be several thousand bits long.

In asynchronous transmission there is at least one start bit and one stop bit for every character of data. If an error occurs during asynchronous transmission, that error may only destroy one character of data because each character is synchronized

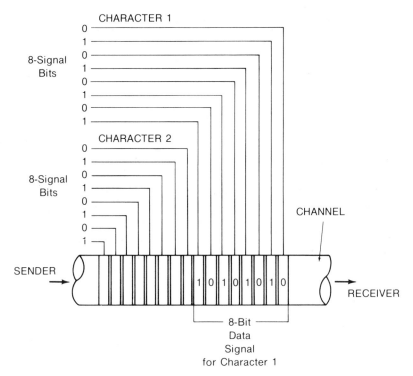

Figure 2-4: Synchronous Transmission of an 8-Bit
USASCII Code in Serial Mode

with its own start and stop bits. On the other hand, that same error in synchronous transmission would probably destroy the entire message block by breaking the synchronization.

The modems and related equipment for synchronous transmission are more expensive than those used for asynchronous transmission because they must be able to synchronize between themselves.

In summary, synchronous and asynchronous data transmission modes are differentiated by the fact that in asynchronous data transmission, each character is transmitted as a totally independent entity with its own start and stop bits to inform the receiving device that the character is beginning and ending. In synchronous transmission, whole blocks of data are transmitted as units after the transmitter and the receiver have been synchronized.

Isochronous transmission is a third technique; it combines the elements of both synchronous and asynchronous data transmission. In isochronous transmission, as in asynchronous, each character is required to have both a start bit and a stop bit. However, as in synchronous data transmission, the transmitter and the receiver are synchronized. The synchronization time interval between successive bits is specified to be an integral multiple of the length of one code bit. That is, all periods of no transmission consist of one or more 1-character time intervals. This common timing allows higher precision between the transmitting and the receiving equipment than could be achieved using asynchronous techniques only.

Figure 2-5 illustrates the relationships and differences between asynchronous, synchronous, and isochronous transmission. In asynchronous transmission, there is no determination of the spacing between individual characters (indefinite time). This requires that both the sending and receiving equipment have clocks to determine the length of a bit, and the receiver must have special recognition circuitry to determine the beginning and end of a character. With synchronous transmission, the clocking signal is used to synchronize the receiver to the sender before a long, multicharacter block of data is transmitted. In isochronous transmission, the clocking is supplied by the sending modem, and the receiving modem synchronizes to it for short periods. Each character begins on some multiple of the length of the bit element.

The primary reason for using isochronous transmission in preference to asynchronous transmission is speed. In practice, asynchronous data transmission is generally limited to 1,800 bits per second by the timing precision of the sending and receiving equipment. By contrast, isochronous data transmission can achieve data transmission rates as high as 9,600 bits per second. Synchronous data communications may be even more rapid than isochronous.

In summary, the modes of transmission are serial (bit-by-bit down the line) and parallel (all bits of a character sent simultaneously). Further, serial transmission (which is most prevalent) can be divided into:

- Asynchronous transmission, in which the data bits of a character are sent independent of the timing of any other character and are preceded by a start bit and followed by a stop bit.

- Synchronous transmission, in which the sending and receiving units are synchronized, and then a stream of many thousands of data signal bits is sent.

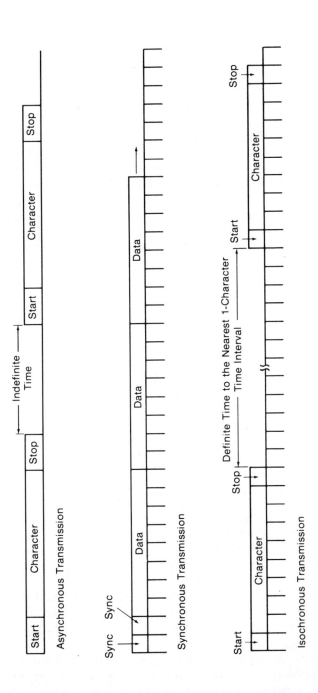

Figure 2-5: Comparison of Transmission Modes

- Isochronous transmission, in which each character still has a start and stop bit but the sending and receiving equipment are synchronized for the length of each timing unit.

Coding Terminology and Structure

A character is a symbol that has a common, constant meaning for some group of people. A character might be the letter A or B, or it might be a number such as 1 or 2. Characters may also be special symbols such as & or ?. Characters in data communications, as in computer systems, are represented by groups of bits. The various groups of bits that represent the set of characters that are the "alphabet" of any given system are called a "coding system," or simply a "code." This section will discuss some of the codes used in data communications.

A byte is a group of consecutive bits that are treated as a unit or character. One byte is normally comprised of 8 bits and it usually represents one character. However, in data communications some codes in regular use utilize 5, 6, 7, 8, 10, or 11 bits to represent a character. These differences in the number of bits per character arise because the codes have different numbers of characters to represent and different provisions for error checking.

Coding is the representation of one set of symbols by another set of symbols. For example, representation of the character A by a group of 7 bits (say, 1000001) is an example of coding.

Many codes use more bits in each character group than is absolutely necessary for unique representation. Many codes also include characters that are not strictly a part of the meaning of any message. The extra bits are used to detect errors, and the extra characters have many uses in error detection or other control functions. These two uses are discussed further in Chapter 4, Network Configuration Concepts and Control Techniques, and in Chapter 5, Error Detection and Correction.

As we have seen, information in data communications is normally transmitted serially over a transmission line or channel. Codes for representing the information vary in relation to both the number of bits used to define a single character and in the assignment of bit patterns to each particular character. For example, the bit group 1000001 may represent the character A in one coding scheme (USASCII), but the bit group 11000 may represent the character A in some other code configuration. In most cases, both the sending and receiving equipment must be engineered or programmed to transmit and receive the code that the data communication system is using. It is this equipment that determines whether one or two stop bits are utilized during asynchronous transmission.

It may be desirable to have a code that detects and/or corrects errors during transmission. Such codes are available and are discussed in Chapter 5. The remainder of this section is devoted to a discussion of various common codes of data communications.

The United States of America Standard Code for Information Interchange (USASCII). This is an 8-bit code with 128 valid character configurations ($2^7 = 128$ characters; the eighth bit is a parity check). USASCII code is the basic standard code for which most communication equipment is designed. This code was born out of earlier attempts to standardize on one code for data transmission. In 1963, the first standardized code appeared. It was called American Standard Code for Information Interchange (ASCII63). A subsequent version appeared in 1965 (ASCII65). Finally, in 1968, USASCII was accepted as a U.S. standard and has been widely adopted nationally. USASCII is 10 or 11 bits per character in asynchronous transmission, utilizing 1 start bit, 7 data bits, 1 parity bit, and either 1 or 2 stop bits. Figure 2-6, USASCII Code Configuration, shows all the valid code configurations.

The Baudot code is a 5-bit code, dating back to the 19th century. It is used mainly by older Teletype equipment. The inventor, Baudot, also gave his name in shortened form to the unit "baud." Figure 2-7 shows how various characters might be represented if they were punched into paper tape, in which the O's are holes and the blanks are "no-holes."

There are 32 different* possible combinations of 1's and 0's in any 5-bit code. This is not enough different combinations to represent the 26 alphabetic characters, the 10 numeric characters, and the several special characters that must be represented. Therefore, Baudot code uses a trick similar to the upper and lower case shift on a typewriter. Two special characters, called "letters" (11111) and "figures" (11011) are used. When one of these characters is used it applies to all codes that follow until the other letter or figure code occurs. Thus, "letters" or "figures" has the effect of a sixth bit in each of the remaining 30-bit groups in

---

*In order to determine the number of possible combinations in any binary coded configuration, raise 2 to the power equal to the number of bits used to represent one character ($2^5 = 32$).

b7 b6 b5 / Bits b4b3b2b1 Col Row	0 0 0 / 0	0 0 1 / 1	0 1 0 / 2	0 1 1 / 3	1 0 0 / 4	1 0 1 / 5	1 1 0 / 6	1 1 1 / 7
0 0 0 0 0	NUL	DLE	SP	0	@	P	'	p
0 0 0 1 1	SOH	DC1	!	1	A	Q	a	q
0 0 1 0 2	STX	DC2	''	2	B	R	b	r
0 0 1 1 3	ETX	DC3	#	3	C	S	c	s
0 1 0 0 4	EOT	DC4	$	4	D	T	d	t
0 1 0 1 5	ENQ	NAK	%	5	E	U	e	u
0 1 1 0 6	ACK	SYN	&	6	F	V	f	v
0 1 1 1 7	BEL	ETB	'	7	G	W	g	w
1 0 0 0 8	BS	CAN	(	8	H	X	h	x
1 0 0 1 9	HT	EM	)	9	I	Y	i	y
1 0 1 0 10	LF	SUB	*	:	J	Z	j	z
1 0 1 1 11	VT	ESC	+	;	K	[	k	{
1 1 0 0 12	FF	FS	comma ,	<	L	\	l	¦
1 1 0 1 13	CR	GS	-	=	M	]	m	}
1 1 1 0 14	SO	RS	.	>	N	⌒	n	~
1 1 1 1 15	SI	US	/	?	O	—	o	DEL

NUL	— Null	SI	—	Shift In
SOH	— Start of Heading	DLE	—	Data Link Escape
STX	— Start of Text	DC-1 to 4	—	Device Control
ETX	— End of Text	NAK	—	Negative Ack.
EOT	— End of Transmission	SYN	—	Synchronous Idle
ENQ	— Enquiry	ETB	—	End of Trans Block
ACK	— Acknowledge	CAN	—	Cancel
BEL	— Bell	EM	—	End of Medium
BS	— Backspace	SUB	—	Substitute
HT	— Horizontal Tab	ESC	—	Escape
LF	— Line Feed	FS	—	File Separator
VT	— Vertical Tab	GS	—	Group Separator
FF	— Form Feed	RS	—	Record Separator
CR	— Carriage Return	US	—	Unit Separator
SO	— Shift Out	DEL	—	Delete (Rubout)

Figure 2-6: USASCII Code Configurations

Ltrs	A	B	C	Figs	—	?	:
○	○	○		○	○	○	
○	○		○	○	○		○
○			○				○
○		○	○	○		○	○
○		○		○		○	

Figure 2-7:   Baudot Code on Punched Paper Tape

the code. Figure 2-7 shows that when the "letters" character is transmitted, all the characters that follow it are considered differently from the same 5-bit code when the "figures" character precedes them. Using the letters and figures shift characters increases the available number of different code configurations to 62. Since three codes plus blank are the same in either shift, Baudot code really has 58 different characters (this count includes the letters and figures shifts) plus blank. Figure 2-8 depicts the 5-bit code configuration for all the alphabetic, numeric, and special characters.

Baudot code, when used in asynchronous transmission, has 5 bits per character (no parity bit), 1 start bit, and a long stop bit equal to 1.42 bit times. This means that a Baudot code character uses 7.42 (1 + 5 + 1.42) bit times during transmission from a sending device to a receiving device. The start and stop bits are used for synchronizing the character during transmission.

There is also a closely related code to Baudot, called Pseudo-Baudot. Pseudo-Baudot is identical to Baudot code except it uses a stop bit equal to 1.5 bit times; therefore, it occupies 7.50 bit times during transmission.

Data Interchange Code is another code that is used on newer Teletype equipment. The Teletype version of the Data Interchange Code is an 8-bit code that uses 7 bits to represent the characters and 1 bit for parity. Data Interchange Code has 128 valid character combinations ($2^7 = 128$ characters). Data Interchange Code is 11 bits per character in transmission, with 1 start bit, 7 data bits, 1 parity bit, and 2 stop bits in asynchronous transmission. Data Interchange Code is primarily used on the slower speed sub-voice grade lines utilizing Teletype equipment manufactured since the early 1960s.

The 4-of-8 Code is an IBM code. It uses only 4 of the 8 bits of data because the only valid combinations that are recognized are

Lower Case	Upper Case	1 byte of 5 bits				
		1	2	3	4	5
A	—	•	•			
B	?	•			•	•
C	:		•	•	•	
D	$	•			•	
E	3	•				
F	!	•		•	•	
G	&		•		•	•
H	£			•		•
I	8		•	•		
J	'	•	•		•	
K	(	•	•	•	•	
L	)		•			•
M	.			•	•	•
N	,			•	•	
O	9				•	•
P	0		•	•		•
Q	1	•	•	•		•
R	4		•		•	
S	Bell	•		•		
T	5					•
U	7	•	•	•		
V	;		•	•	•	•
W	2	•	•			•
X	/	•		•	•	•
Y	6	•		•		•
Z	"	•				•
Letters (shift to lower case)		•	•	•	•	•
Figures (shift to upper case)		•	•		•	•
Space				•		
Carriage return					•	
Line feed			•			
Blank						
• represents a 1 Blank represents a 0						

Figure 2-8: Baudot Code Configuration

those in which exactly 4 of the 8 bits are in a 1 configuration and the other four are in an 0 configuration. For example, this would disallow as a valid data bit combination the following configuration: 10101110. The purpose in allowing only configurations that have four 1's and four 0's is that an accuracy check can be accomplished by ensuring that there are always four 1's and four 0's in the received data character. The penalty paid for doing this is that there are only 70 valid characters, instead of 256 ($2^8 = 256$ characters). The 4-of-8 code is 10 bits per character in asynchronous transmission, with 1 start bit, 8 data bits, and 1 stop bit. This code detects errors better than the single parity bit of USASCII or the other codes that use a single parity bit. The 4-of-8 Code is used primarily on high-speed voice grade lines.

The Binary Coded Decimal (BCD) Code is an extension of the older tab-card-oriented Hollerith code. It is a 6-bit code and has 64 valid character combinations ($2^6 = 64$ characters). Binary Coded Decimal Code is 9 bits per character in asynchronous transmission, with 1 start bit, 6 data bits, 1 parity bit, and 1 stop bit. This code is used primarily on low-speed lines.

The Extended Binary Coded Decimal Interchange Code (EBCDIC) is IBM's System 360/370 code. EBCDIC has 256 valid character combinations ($2^8 = 256$ characters). This code has 11 bits per character in asynchronous transmission utilizing 1 start bit, 8 data bits, 1 parity bit, and 1 stop bit.

The codes described above are the basic ones utilized in data transmission. These codes can be used in asynchronous, synchronous, or isochronous transmission. Figure 2-9 summarizes the various start bit, data bits, parity bits, stop bits, and the total

Code	Start Bit	Data Bits	Parity Bit	Stop Bit Time	Total Bit Times per Transmitted Character
USASCII	1	7	1	1 or 2	10 or 11
Baudot	1	5	none	1.42	7.42
Pseudo-Baudot	1	5	none	1.50	7.50
Data Interchange	1	7	1	2	11
4-of-8	1	8	none	1	10
Binary Coded Decimal	1	6	1	1	9
EBCDIC	1	8	1	1	11

Figure 2-9: Code Configurations for Asynchronous Transmission

bits in transmission for the seven different code configurations that have been discussed.

From this discussion, it is obvious that the number of data bits used to represent a letter, number, or symbol is dependent upon the code used. Each code may have a different number of possible characters; for example, Baudot has 57, and EBCDIC has 256. Each code has different characteristics as far as its inherent ability to detect alterations in bit patterns (caused by transmission problems) is concerned. Baudot has no error detection, USASCII has one parity bit for error detection, and 4-of-8 has a unique method of error detection. The speed of the data transmission line required depends upon how many bits are to be transmitted during a given period of time. Thus, the Data Interchange Code, which takes 11 bits per character, is not as efficient as the BCD code, which takes 9 bits per character. In other words, 100 characters using the BCD code would require 900 total bits in asynchronous transmission, while the same 100 characters using the Data Interchange Code would require 1,100 bits in asynchronous transmission.

When designing an information system, it is necessary to consider the different forms and encodings of information that may be needed. As a rule, when information appears in different media (for example, punched cards and punched paper tape) the information may also be in a different code. This may require either software to perform a code conversion or a specially designed hardware code conversion unit. Some code configurations are dependent on the equipment being used. For example, to accommodate users with different requirements, the Teletype Model 33 Terminal uses the Data Interchange Code, while the Teletype Model 32 Terminal uses Baudot.

Types of Transmission

A channel is a path for electrical transmission between two or more points. A channel is also referred to as a line, a circuit, a link, a facility, or a path. The most common terminology is a channel, circuit, line, or a link.* When we use these terms we are usually talking about the telephone wires (or their equivalents) that are present everywhere. Basically, the telephone lines are analog channels that will pass alternating current, but not direct current. Analog signals are continuous rather than "on-off." Digi-

---

*"Link" is generally reserved for one point-to-point line (See Figure 1-3).

tal signals, on the other hand, are "on-off." If an analog channel is said to carry digital data, it is actually carrying analog representations of the digital data in some form. Lines are classified by the types of media through which they pass, for example copper wire, cables, atmosphere, coaxial cable, and the like.

The available methods of transmission are simplex, half duplex (HDX), and full duplex (FDX), as shown in Figure 2-10. In simplex transmission, information is transmitted in one direction only, and the roles of transmitter and receiver are fixed. A doorbell is an example of simplex transmission. Simplex transmission is not utilized in conventional data communication networks. In HDX transmission, one station transmits information to another and when that transmission has been completed, the other station can respond. In other words, HDX transmission allows transmission in both directions but only in one direction at a time. A polite conversation in which neither participant interrupts the other is an example of HDX transmission. In FDX transmission, both stations can transmit and receive simultaneously. Information can flow on the lines in both directions at the same time. An argument in which participants speak at the same time is an example of FDX transmission. Data communication systems that use the regular telephone dial-up network usually transmit in HDX. In order to transmit in FDX the data communication user usually has private or leased voice grade lines.

The types of circuits normally available are "two-wire" circuits and "four-wire" circuits. Two-wire circuits are just that; two wires connecting one station to another station. Since it takes two wires to send a signal in one direction at a time, two-wire circuits can handle only simplex or HDX. With four-wire circuits, the user has four wires connecting one station to another station, thus making FDX possible.

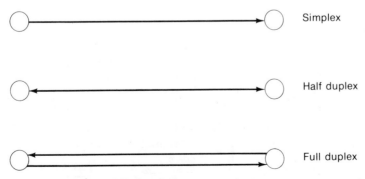

Figure 2-10: Methods of Transmission

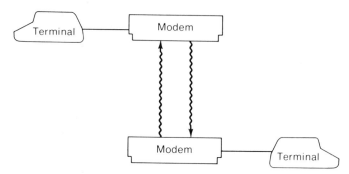

Figure 2-11:   Two-Wire Connection

Figure 2-11 shows a two-wire connection, and Figure 2-12 a four-wire connection. Two-wire and four-wire connections represent a physical configuration, whereas HDX and FDX represent a method of transmission. In order to transmit FDX (simultaneous transmission in both directions), the terminals and software at each end of the transmission must be designed to handle simultaneous communications.

Two-wire circuits have a problem of echoes. When people talk on a two-wire circuit, echoes may occur under some conditions. Echoes arise in telephone circuits for the same reason that acoustic echoes occur: there is a reflection of the electrical wave from the far end of the circuit. The telephone company provides echo suppression circuits to stop echoes during voice conversation. An echo suppressor permits transmission in only one, fixed, direction. These echo suppressors open and close on two-wire lines during data transmission. Figure 2-13 shows the steps of the opening and closing of echo suppressors as Terminal A transmits

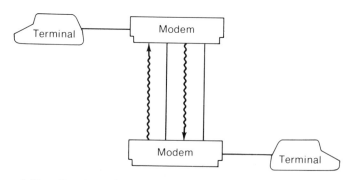

Figure 2-12:   Four-Wire Connection

to Terminal B and Terminal B responds back to Terminal A. As A sends to B, the echo suppressor on one of the two lines closes and the other echo suppressor opens to stop return echoes back to A. When B sends to A, the one echo suppressor closes and the other opens so no echoes will return to B. These echo suppressors cause noise and static on the lines, which may cause errors. They also cause a delay in data transmission. The delay results because B must wait for one echo suppressor to close and the other to open before he can send his answer back to A. This wait is approximately 150 milliseconds. To you and me this might seem trivial, but to a high-speed data communication system, a 150 milliseconds wait is time-consuming and wasteful. Four-wire lines transmit in one direction only on each two-wire segment; therefore, there is no turnaround delay for echo suppressors because the same echo suppressor is always closed.

The telephone company has provided a way to assist data communications by transmitting a tone over the line that closes or disables the echo suppressors. When they are closed there will be echoes, but these do not interfere with the data communications (you will recall that echo suppressors are put on the circuits

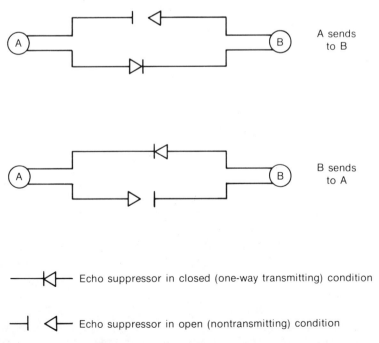

Figure 2-13:   Echo Suppressors on a Two-Wire Line

for human voice transmission). Another solution to this problem is to use a four-wire circuit. Review Figure 2-12 and you will see that a four-wire circuit is nothing more than two, two-wire lines. Now look at Figure 2-13 again and assume that A is talking to B on the upper two-wire line of that figure and B is talking to A on the lower two-wire line. In this case, the proper echo suppressors would always remain closed because there is transmission in one direction on one two-wire line and transmission in the other direction on the other two-wire line. Figure 2-14 summarizes facts about two- and four-wire circuits in relation to HDX and FDX transmission.

	Terminal Method of Operation	Property of the Communications Line (Channel)
Half Duplex (two-wire)	Terminals are not capable of simultaneous sending and receiving, regardless of the communication lines.	Provides one transmission path, which may be used in either direction but not in both directions at the same time.
Full Duplex (four-wire)	Terminals are capable of sending and receiving simultaneously.	Provides two independent transmission paths, one in each direction.

Figure 2-14:   Types of Transmission

Characteristics of Transmission Media

There are two general categories of electrical current: direct current and alternating current. Direct current (DC) travels in only one direction in a circuit, while alternating current (AC) travels first in one direction (+) and then in the other direction (−). The frequency of a continuous AC wave is the number of times per second that the wave makes a complete cycle from 0 to its maximum positive value, then through to its maximum negative value, and back to 0. If the AC is generated by a constant speed rotating generator, or a circuit that simulates such a device, each full cycle is one complete sine wave. The sine wave is described by the mathematical expression $C \cdot \mathrm{Sin}\pi Kt$, where C is the amplitude (or maximum positive value) and $T = 1/K$ is the period. Figure 2-15, shows three different sine waves. The top

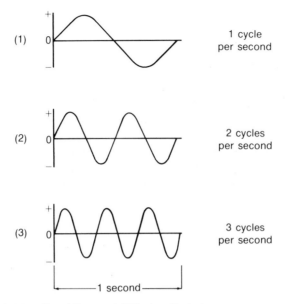

Figure 2-15:   Sine Waves of Differing Periods

configuration (1) shows one complete cycle; the second configuration (2) shows two complete cycles; and the third configuration (3) shows three complete cycles.* Almost all transmission over telephone lines utilizes the AC wave form known as a sine wave.

Some transmission is performed using DC where the signals are represented as voltages of DC electricity being transmitted down the wire. These DC signals are never sent over wire pairs of more than a few thousand feet in length nor at speeds more than 300 bits per second. It is somewhat impractical to send DC signals at speeds much greater than 300 bps because the distortion of the signal makes it difficult for the receiving device to interpret the signal. To exceed a distance of 3 to 9 miles, a repeater is required (this regenerates the signal to its original strength) every few thousand feet (See Figure 2-16).

Frequency is measured in cycles per second; also often expressed as Hertz (Hz). The higher the frequency of a wave, the more it reverses direction each second. Sine waves can be produced with frequencies from as low as a few cycles per second to

---

*For all three, the value of C is the same, but $T_{(1)} = 2T_{(2)} = 3T_{(3)}$. That is, the period of (2) is one half that of (1), and the period of (3) is one third that of (1).

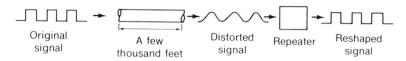

Figure 2-16:   Use of a Repeater in Long DC Lines

trillions or more Hz. Because of this extremely wide range, frequencies are usually divided into subranges or bands. Each band is a range of frequencies that have similar properties and are used in similar ways in electronic applications. Figure 2-17, shows the voice/audio frequency spectrum that goes from 20 Hz up to 20,000 Hz. This portion of the frequency spectrum spans the frequency range of an average person's hearing. Within that range is the voice grade telephone band. The telephone channels over which data communication networks transmit have a bandwidth of 3,000 Hz. Basically, a telephone channel ranges from 300 cycles per second to 3,300 cycles per second. (Figure 2-15 depicts 1 cycle to 3 cycles per second.)

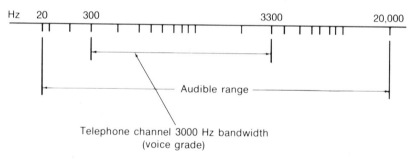

Figure 2-17:   Voice/Audio Frequency Spectrum

The major difference between the grades of available telephone channels is the bandwidth. For example, as you will see in a later chapter, two of the available lines are low-speed lines and voice grade lines. Voice grade lines have a bandwidth of 3,000 cycles per second and low-speed lines have a much narrower bandwidth. Therefore data communication equipment cannot transmit over low-speed lines at as high a speed (number of bits per second transmitted) as it can over voice grade lines.

The rest of this section discusses the various types of transmission media in use today. Some of these media, such as microwave links, operate at a very high frequency, while coaxial cables operate at a lower frequency and open wire pairs operate at a very low frequency.

Open wire pairs   are most familiar in rural areas. They are copper or copper-colored wires suspended by insulated crossarms on telephone poles. These open wire pairs do not have any insulated coating; they are spaced about 8 to 12 inches apart. Most of the open wire pairs have now been replaced by cables and they are fast becoming part of the romantic past.

Wire cables   have replaced open wire pairs. Wire cables are insulated and therefore can be brought closer together and packed into one large cable. These wire cables are twisted in pairs to minimize the electromagnetic interference between one pair and another when they are packed into a large cable. Many hundreds of wire pairs may be grouped together in a large cable. These cables are laid under the streets of cities and are the primary path (the local loop) between a subscriber's premises and the telephone company's local central exchange office. Each of these wire pairs (a two-wire circuit) is capable of carrying one voice grade telephone channel.

A coaxial cable   can transmit at much higher frequencies than a wire pair. It consists of a hollow copper cylinder, or other cylindrical conductor, surrounding a single wire conductor. The space between the hollow copper cylinder and the inner conductor is filled with an insulator, which separates the outer shell and the inner conductor. These insulators are spaced every few inches or so. Figure 2-18 shows the construction of a coaxial cable. These coaxial cables can be bundled into a large cable that contains 20 coaxial cables and can handle up to 18,740 telephone calls at the same time. Coaxial cables have very little distortion, cross talk, or signal loss and therefore are a better transmission medium than either twisted wire cable or open wire cable.

Microwave transmission   is the main competition for coaxial cables because it can carry many thousands of voice channels at once and it does not require the laying of a cable. Microwave transmission is relayed through the atmosphere between microwave towers that are usually spaced 25 to 30 miles apart. Microwave is a line-of-sight transmission method—the receiver

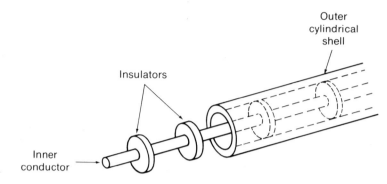

Figure 2-18:   Coaxial Cable

must be able to "see" the transmitter. Each tower picks up the transmitted signal from the previous tower, amplifies it, and retransmits it to the next microwave tower. A typical antenna for a microwave tower is about 10 feet across, although over shorter distances these antennas may be smaller.

Submarine cables are coaxial cables with a larger spacing between the inner conductor and outer cylindrical shell; they are specially constructed to withstand undersea environments. One of these submarine cables can handle over 700 simultaneous transmissions. Today, submarine cables are being supplanted by communication satellites, which are less costly per channel.

Communication satellites provide a special form of microwave relay transmission. The satellite is nothing more than a very high microwave tower placed miles above the earth usually over the equator. Thus, it can relay signals over longer distances than is possible on earth because the curvature of the earth, mountains, and other obstacles block line-of-sight microwave transmission between land-based microwave towers. Satellites can handle many thousands of voice grade transmissions simultaneously. Currently, commercial satellites are put into a very high orbit (22,300 miles) so they travel at the same speed as the rotation of the earth. This "synchronous" orbit causes the satellite to appear to hang stationary over a particular spot on the earth. In this way, only three satellites placed in high orbit relay data communications completely around the earth except for the extreme polar regions (see Figure 2-19).

Lasers and optical fibers are two of the new technologies for data transmission now being tested for practical use. A laser generates

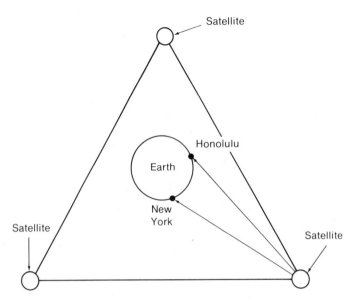

Figure 2-19:   Satellites

a very high frequency, coherent beam of light which is capable of transmitting 100,000 times as much information as today's microwave links. Optical fibers (thin filaments of glass or glass-like materials) are an experimental transmission medium that shows special promise when coupled with lasers.

High-frequency radio   is another lesser used form of data transmission. High-frequency radio rarely forms part of a computer data communication system except for transmitting (at a very low transmission rate) from remote parts of the world. This type of data communications is seldom used because the data error rate is extremely high and elaborate means of error detection and correction are required. One form of this transmission is called tropospheric scatter. Transmission by this method is limited to less than 600 miles using radio waves. One drawback to this method of transmission is that it requires very large antennas, on the order of 60 feet high and 120 feet long.

Wave guides   are also used experimentally for data communications. A wave guide is a metal tube or a rectangular box-shaped tube down which radio waves of very high frequency travel. The radio waves bounce around inside the tubes. The circular tubes are approximately 2 inches in diameter. Wave guides are rarely used for distances over a few thousand feet, although they have a capacity of 100,000 or more voice grade lines. Wave guides are used to connect the antennas of radars and microwave transmission towers to the transmitting equipment located nearby.

Circuit Concepts

Four terms that are important to the data communication specialist are "baud," "dibit," "unipolar," and "bipolar."

A baud is a unit of telegraph signaling speed that is found by taking the reciprocal of the length (in seconds) of the shortest pulse used to create a character. The length of a pulse of Baudot code used with a 60-word-per-minute Teletypewriter is 0.022 seconds; therefore, the baud rate is 1/0.022 = 45.45 baud (the plural of baud is baud). "Baud" and "bits per second" are not synonyms, but most data communication practitioners use them interchangeably. A bit is a unit of information. A baud is a unit of signaling speed. Bit rate and baud rate coincide only when a code is used in which all bits are of equal length. Because this is true in most cases, one can use these two words interchangeably and they will be understood by data communication people. In transmission technology, one pulse is generally equal to a single-bit state, e.g., a 1,200-baud circuit generally implies a 1,200-bit-per-second transfer rate. Because the use of the word "baud" has proved confusing we will not use it in this book. We will use "bits per second" exclusively.

One simple method of increasing the rate of bits transmitted through a line is to combine pairs of adjacent bits into dibits ("di" = two). The objective is to send each dibit as a separate signal element. Because the laws of information theory cannot be repealed, the process of transmitting two bits at the same time involves four (2^2) different signal states. One way to get four states is depicted in Figure 2-20. One signal state would represent 00, another 01, another 10, last one 11. Figure 2-21 shows how, by varying the voltage between 0 and 3 volts, the transmitting device can send 2 bits simultaneously. This figure depicts the original bit stream of 0's and 1's which were represented by −1 volt for a 0

```
0  0   equals   0 volts
0  1   equals   1 volt
1  0   equals   2 volts
1  1   equals   3 volts
```

Figure 2-20: Four States with 0 or 1 Taken Two at a Time

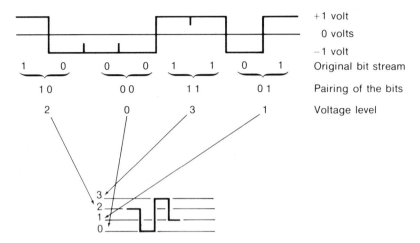

Figure 2-21: Dibits Represented by Different Voltages

and +1 volt for a 1. Below that original bit stream is the bit stream of paired bits (dibits) and each of the four different voltage levels that represents each dibit. In other words, whenever the receiving device receives a +3 volt signal it will interpret that it has received the pair of bits "11." If the receiving device receives the +1 volt signal it will interpret that it has received a pair of bits "01." Figure 2-22 shows dibits represented by amplitude modulation (amplitude modulation is introduced in the next section).

A USASCII character that is transmitted by serial asynchronous transmission will require 10 or 11 bits during transmission. If the transmission rate is 1,200 bits per second (bps), each of these bits will require 833 microseconds (μsec) or 8,330 μsec per character (using 10-bit characters). By using the dibit configurations, at the same rate as before, the same character will require only one half that time (4,165 μsec) because each of the 833 μsec signals will be transmitting two bits instead of one. This technique helps to squeeze higher bit rates out of telephone lines that have basic limitations on the number of state changes per second that can be handled. In particular, 9,600-bps transmissions over voice grade lines take advantage of "tricks" like this. More expensive transmitting and receiving equipment is required, however, and usually the conditioning of lines to reduce errors.

Two types of electrical line conditions should be understood: unipolar and bipolar signals. Unipolar describes a line condition

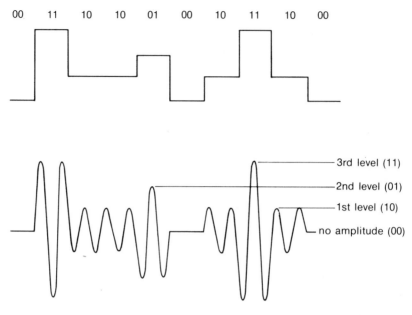

Figure 2-22:   Dibits Represented by Amplitude Modulation

where the signal is switched on and off at fixed or arbitrary
intervals. This method was used on telegraph equipment in its
early days. Turning a light on and off is a unipolar process.
Bipolar describes line conditions where the signal is reversed
intermittently by applying either positive or negative potential to
the line (Figure 2-23). The opposing polarities distinguish the 1
and 0 states. Bipolar transmission is the basic transmission
method used in today's data communication lines.

Modulation

A continuous oscillating voltage of arbitrary amplitude and frequency carries no intelligence. However, if it can be interrupted
or the amplitude altered so it becomes somewhat like a series of
pulses that correspond to some known code (such as USASCII)
then the oscillating signal can carry some intelligence. In data
communications, this continuous oscillating voltage is called a
"carrier signal" or simply a "carrier." Figure 2-24 schematically
illustrates the carrier signal that is always present between two
modems. The carrier signal can be altered in many ways. The
process of changing some characteristic (i.e., amplitude, fre-

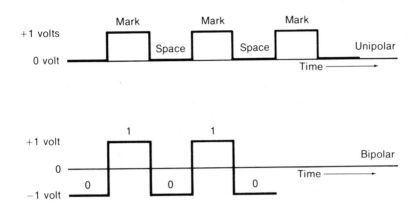

Figure 2-23: Unipolar and Bipolar Signals

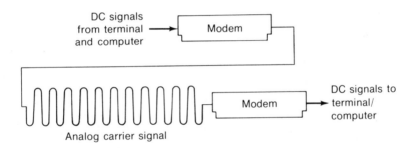

Figure 2-24: Carrier Signal

quency, or phase) of a carrier signal in order to transmit meaningful information using that signal is called modulation. The four most common methods of modulation are amplitude modulation, frequency modulation, phase modulation, and pulse modulation.

The equipment in which modulation is performed is called a modulator. If the modulator causes the amplitude of the carrier signal to vary, the result is called amplitude modulation, and so on for frequency, phase, and pulse modulation. These modulators are, in fact, the modems that are connected to each end of the transmission lines. The name modem is made up of the words *mo*dulator and *dem*odulator. The modem that is transmitting the signal is the modulator because it modulates, or puts some form of intelligence on the carrier wave, whereas the receiving equipment is the demodulator because it demodulates or interprets that signal upon its receipt. In other words, the process of modifying a carrier so it carries a signal that can be interpreted is referred to as modulation and the process of converting it back again so the original intelligence is recovered is called demodulation. The effect here is to take the binary signal (digital signal) from a computer or business machine and modulate it so it will become a continuous signal (analog signal) that can be transmitted over telephone lines or microwave towers, etc. In the demodulation process, the receiving equipment interprets the modulated carrier signal (analog signal) and converts it to a binary signal (digital signal) that will be meaningful to the computer or other business machine.

The remainder of this section will discuss the four most common types of modulation; amplitude, frequency, phase, and pulse code modulation.

In amplitude modulation (AM),   the peak-to-peak voltage of the carrier signal is varied with the intelligence that is to be transmitted. The amplitude modulation shown in Figure 2-25 depicts the peaks at one amplitude representing binary 1's, and the peaks of another amplitude representing binary 0's. Several levels of amplitude modulation are possible; four levels will allow twice as much data to be sent in the same elapsed time (see Figure 2-22). Amplitude modulation is suitable for data transmission and it allows efficient use of the available bandwidth of a voice grade line. However, frequency modulation has the advantage of being less susceptible to noise during transmission than amplitude modulation.

Frequency modulation (FM)   is the most common form of modulation at transmission rates of up to 1,800 bps. In frequency modulation, the carrier signal is modulated to different frequencies. For example, the carrier signal may be modulated back and forth between 1,200 Hz and 2,200 Hz (without affecting amplitude) in response to the binary digital signal. The specific frequencies used depend on the transmitting and receiving equipment utilized. For example, one type of equipment may represent a 0 with 1,200 Hz and 2,200 Hz may represent a 1. Figure 2-26 shows how the carrier signal might be modulated using frequency modulation.

When FM is used to send binary information in bipolar form, it is known as Frequency Shift Keying (FSK). In this system, the carrier signal (let us assume it is operating at 1,700 Hz) is

0   0   1   1   0   1   0   0   0   1   0

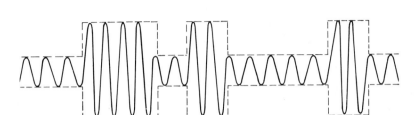

Figure 2-25: Amplitude Modulation

modulated plus or minus 500 Hz to represent a binary 1 or a binary 0. Thus, the frequency 2,200 Hz might be a binary 1 and the frequency 1,200 Hz might be a binary 0. If the transmission rate is 1,200 bps, and a frequency of 2,200 Hz is sent down the line for approximately 833 μsec it will represent a binary 1. When a frequency of 1,200 Hz is transmitted down the line for 833 μsec, the receiving equipment interprets that to mean a binary 0 was transmitted. The FSK technique is generally suitable for low-speed devices such as teleprinters that operate at 1,800 bps and below. FM is less affected by noise on the transmission lines than is AM (remember, FM radio is generally static-free). Therefore, it produces data transmission that is less error prone.

Phase modulation (PM) is now beginning to replace amplitude and frequency modulation for high-speed transmission because it is affected even less by noise than AM or FM. In PM, the phase* of a carrier signal is varied in accordance with the data to be sent. The modems that utilize phase modulation are generally described in terms of the number of phase shifts generated. Phase modulation is generally utilized in equipment that operates at speeds above 2,000 bps. The phase of the transmitted signal is shifted a certain number of degrees in response to the pattern of bits that are to be transmitted. In a two-phase modem (similar to frequency shift keying) the signal is shifted 180° (360 ÷ 2) depending upon whether a binary 1 or 0 is indicated. If there is no shift, then the signal will be interpreted as a series of 1's or 0's.

---

*"Phase" is the measure of the relative time that a sine wave crosses through zero amplitude. See glossary for more details.

0   0   1 ·   1   0   1   0   0   0   1   0

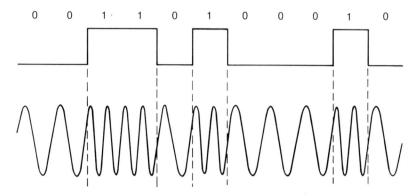

Figure 2-26:   Frequency Modulation (The Cycles per
Second Vary—See Figure 2-15)

Figure 2-27 depicts Differential Phase Shift Keying (DPSK). In this technique, whenever a 180° phase shift is encountered, the receiving equipment assigns a value of binary 0; at all other times, the value of a binary 1 is assumed. Generally, phase modulation equipment operates in four and eight phases, permitting up to twice or three times as many bits to be sent over the line in the same bandwidth in a given time. Twice as much information can be sent using the dibits concept and three times as much information could be transmitted using a tribits concept

1   0   1   1   0   1   1   0

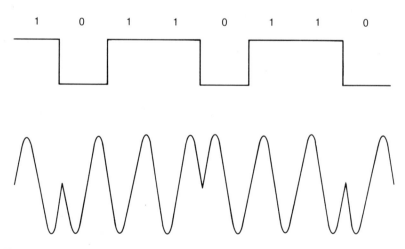

Figure 2-27:   Phase Modulation (The Phase Varies)

(tribits imply having eight different signal levels so the equipment can transmit three bits simultaneously with each signal).

In pulse code modulation (PCM), a coded pulse is used to represent quantitized values of instantaneous samples of the signal wave. The characteristics of pulses within a pulse train may be modified in one of several ways to convey the information. Figure 2-28 shows three different ways in which pulse modulation is utilized. This type of modulation is used in digital transmission.

In pulse code modulation, the information is converted into a stream of bits looking remarkably like computer data. For example, in pulse amplitude modulation (Figure 2-28), the modulating signal is first converted to a series of pulses whose amplitudes correspond to the instantaneous amplitudes of the modulating signal. These amplitude-modulated pulses are then used to amplitude modulate a carrier wave, with the result that the modulated carrier consists of a series of pulses having amplitudes that correspond to the intelligence being transmitted.

Pulse modulation is a digital technique for transmission. This digital transmission technique is the most noise free of the four types of modulation depicted in this section; therefore, to utilize pulse modulation is to minimize the error rate during transmission.

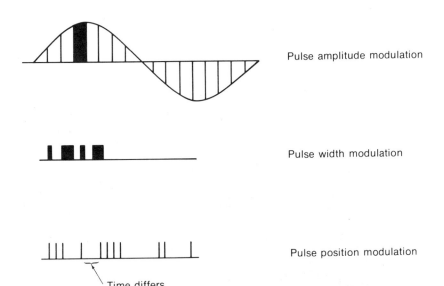

Pulse amplitude modulation

Pulse width modulation

Pulse position modulation

Time differs

Figure 2-28:   Pulse Modulation

In summary, modulation has been described so the systems designer will have a basic understanding of the subject. In reality, the modulation technique is already designed into the modem (modems are discussed in the next chapter). Therefore the network designer does not directly specify the type of modulation except through the modems used in the network.

Questions—Chapter 2

This chapter has concentrated on what happens to data when they are moved from point to point. There are four kinds of questions: True or False, Fill-In, Multiple Choice, and Short Answer. The short answer questions are further subdivided, with Questions 1 through 5 designed to put some of the concepts you have learned to work; try a couple of these, even if they are not assigned to you to be done. These questions test your ability to analyze and then generalize, a key two-step process in designing any kind of system. In solving these problems, you are performing, on a small scale, the same process you will use in solving "real-world" problems. The remaining short-answer questions test for some of the facts you have learned in this chapter. Do them twice; the first time "closed book," the second time "open book"; then compare your answers.

True or False

1.   In parallel transmission, n bits take n cycles to transmit.

2.   Parallel transmission generally requires a higher cost transmission line than serial transmission and parallel transmission also generally requires a high cost receiver.

3.   Synchronous transmission requires start and stop bits for each character.

4.   Synchronous transmission is used for the high-speed transmission of blocks of characters.

5.   Some code configurations are dependent on the equipment being used.

6.   "Baud" and "bits per second" are synonyms.

7.   Hz or Hertz refers to the amplitude of a wave.

8.   For a dibit of information to be transmitted at one time requires that four signal states could be transmitted.

9.   Bipolar transmission is the basic transmission method used in today's data communication lines.

10.   Data Interchange Code is an 8-bit code that includes one parity bit.

Fill-in

1.   The internal transfer of data within modern computers is done in ——— mode.

2.   There is no fixed determination of the time intervals between the transmission of individual characters in ——— transmission.

3.   USASCII stands for ——— ——— — ——— ——— ——— — ——— ——— .

4.   ——— signals are continuous rather than on-off.

5.   In ——— transmission, information is transmitted in one direction and the roles of transmitter and receiver are fixed.

6.   ——— ——— allows transmission of information in two directions, but in only one direction at a time.

7.   ——— is a line-of-sight transmission method.

8.   Transmission lines with _____ current require repeaters for transmissions exceeding about 3 to 9 miles.

9.   The process of changing some characteristic of a carrier signal in order to transmit meaningful information is called ——— .

10.   A ——— ——— consists of a hollow copper cylinder, or other cylindrical conductor, surrounding a single wire conductor with an insulator that separates the outer shell from the inner conductor.

Multiple Choice

1.   Which of the following modes of transmission uses a start and one or more stop bits each time a character of data is transmitted?

a) Asynchronous
b) Isochronous
c) Synchronous
d) A and B above
e) None of the above

2. Which of the following modes of transmission are parallel?
a) Asynchronous
b) Isochronous
c) Synchronous
d) A and B above
e) None of the above

3. The type of modulation where the peak to peak voltage of the carrier signal is varied is called:
a) Amplitude modulation
b) Frequency modulation
c) Phase modulation
d) Pulse modulation
e) None of the above

4. The type of modulation that is usually most affected by noise is called
a) Amplitude modulation
b) Frequency modulation
c) Phase modulation
d) Pulse modulation
e) All are equally affected

5. The type of modulation that is usually least affected by noise is called
a) Amplitude modulation
b) Frequency modulation
c) Phase modulation
d) Pulse modulation
e) All are equally affected

6. An 8-bit code that can represent 256 characters is
a) Baudot
b) BCD
c) EBCDIC
d) USASCII
e) 4-of-8 code

7. An 8-bit code that uses 1 parity bit and can represent 128 characters is
a) Baudot

b) BCD
c) EBCDIC
d) USASCII
e) 4-of-8

8. A 6-bit code that can represent 64 characters is
   a) Baudot
   b) BCD
   c) EBCDIC
   d) USASCII
   e) 4-of-8

9. An 8-bit code that does not use a parity bit for data transmission but can represent only 70 characters is
   a) Baudot
   b) BCD
   c) EBCDIC
   d) USASCII
   e) 4-of-8

10. Four wires are necessary for which of the following types of transmission:
    a) Full-duplex
    b) Half-duplex
    c) Simplex
    d) A and B above
    e) B and C above

Short answer

1. As any radio listener knows, FM is less susceptible to noise (static) than is AM. Can you think of a reason why?

2. If there were a 3 out of 7 code, it could represent $C_3^7 = 7!/(3!4!) = 35$ distinct symbols, out of the $2^7 = 128$ bit combinations. We could then say that the 3 out of 7 code is $35/128 = 0.273$ or 27.3% "efficient." How efficient would a 5 out of 10 code be?

3. If you experiment with a Touch-Tone telephone and listen carefully (and if you are not tone-deaf) you will discover that each key generates two tones. One of the tones is the same for all keys in the same row; the other is the same for all keys in the same column. Label the "row tones" a,b,c,d from lowest to highest pitch, and the "column tones" x,y,z, from the lowest to the highest pitch and make an equivalence table between the digits 0 through 9, * and #, and

the tone code letter pairs. In the terminology of bits, dibits, tribits, etc., when you press one key what are you generating?

4. Parallel transmission is almost always used over short distances, as in cables between computer components, whereas serial transmission is used over the long distances. Can you think of the main reason why?

5. For some data communication systems that you can observe in use, determine the transmission (parallel/serial, synchronous/asynchronous), the code used, whether it is half duplex or full duplex and the transmission medium.

6. What is a start bit in asynchronous transmission?

7. In what type of transmission are whole blocks of data transmitted as units of data between the transmitting terminal and the receiving station?

8. Which code has 11 bits per character in asynchronous transmission?

9. What type of circuits have problems with echoes?

10. Which would require the most bits to represent a character during transmission, a USASCII asynchronous character or a Baudot asychronous character?

3

Data Communication Hardware

The previous chapter discussed methods of communicating data; this chapter describes equipment to accomplish those methods. A large and ever-growing array of equipment is available to data communication networks for transmitting and receiving data as well as for performing the basic tasks of handling the data as they flow through the network. Hardware selection takes into account the network requirements, which, in turn, are derived from the basic business requirements of the organization that will utilize the data communication network. This chapter discusses data communication hardware, starting with central computers, and front-end communication processors, on to modems, multiplexers/concentrators, and individual terminals through which the users of the system enter data and receive the information they have requested.

Central Computers

The suitability of a computer to serve as the central computer for an on-line real time data communication system depends upon both its own capabilities and the capabilities of other hardware attached to it. Many computers on the market today can be used for on-line real time data communication networks, provided that the hardware attached can handle the tasks for which the central computer is inefficient. In other words, the characteristics that make a computer suitable for data communications do not necessarily make it good for "number crunching." In particular, data communication work involves many short periods of activity to service a single arriving or departing character or message. A computer whose hardware or software make this kind of operation clumsy or time consuming will not perform well in the data communication environment. To make such a machine effective, auxiliary hardware is required.

Data communication configurations fall into three categories, dependent upon the interface between the data communication network and the processing functions of the central computer. The *first* of these three categories is a stand-alone computer configuration. In this configuration, the computer is designed to handle a specific set of communication facilities and terminals. The circuitry to handle this is built directly into the computer. In other words, the computer's architecture is designed so that it can interact in a real time mode. Figure 3-1 shows a stand-alone communications configuration in which the central computer is able to handle all the communication protocols. This type of computer is a stored program computer with communication as well as computing capabilities. This computer is typically used in a mode where the emphasis is on communication rather than on data processing. It is often seen in the manufacturing environment for process control, and in areas where the user queries a data base on the status of a certain product, inventory level, or the like. This field is dominated by minicomputers that have been developed and programmed for special purpose processing and communication functions.

The *second* category is a general purpose computer. General purpose computers normally are not designed with built-in communication interface hardware. This type of computer can handle a small data communication network, although, as the number of terminals increases, the general purpose computer begins to reach a point where severe degradation of performance sets in because of its inefficient handling of the communication part of its job. This type of application is exemplified by the organization that has a large general purpose computer that is used for both batch processing and some intra-organizational time sharing functions. These time sharing functions might involve a few terminals that programmers utilize for on-line programming capability. This is usually a limited operation because the general purpose computer is primarily utilized for batch processing. If too many time sharing terminals are connected, the central computer becomes overburdened with the communication functions.

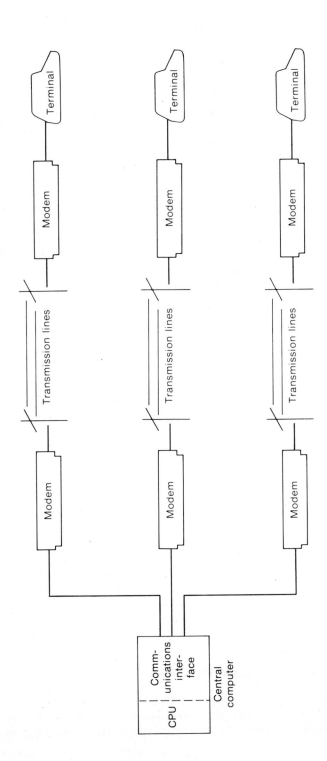

Figure 3-1:   Stand-Alone Communication Configuration

The *third* category, which is becoming more common today, is the front-end configuration. In this case, a large general purpose computer is used for both data communications and maybe some batch processing, but with greater emphasis on the on-line real time data communication portion of the system. In this configuration, there is a distinct division of labor between the front-end module and the general purpose computer.

The front-end module can take two forms. The first is that of a nonprogrammable, hard-wired, communication control unit that is designed by the computer manufacturer to adapt specific line and terminal characteristics to the computer. The second form is that of a front-end communication processor that is programmable and can handle some or all the input/output activity as well as performing some processing. Figure 3-2 depicts a front-end/central computer configuration.

Such a configuration is primarily employed in situations where the input/output and computing processing requirements are very large and where rapid response time is of the essence. This is the type of configuration used in all large data communication networks. The characteristics of both the nonprogrammable communication control unit and the programmable front-end communication processor will be discussed in the next section.

The distinctions between the three data communication categories discussed here are sometimes hazy because of the overlap between these categories in the current networking state of the art. For example, the stand-alone communication configuration (first category) can sometimes be utilized as the front-end communication processor. The general purpose computer configuration (second category) is usually the main computer to the front-end communication processor (third category). Therefore, the reader should recognize that these three categories are sometimes intermixed into a hybrid category that involves two or even all three of these categories in a major data communication network. Figure 3-3 depicts a point-to-point network and shows a central computer, front-end communication processor, modems, and terminals. The following sections of this chapter will discuss these specific pieces of hardware as well as multiplexers/concentrators.

Front-End Communication Processors

As mentioned above, these devices are of two types, nonprogrammable or fully programmable. Nonprogrammable units are also known as communication controllers or transmission control

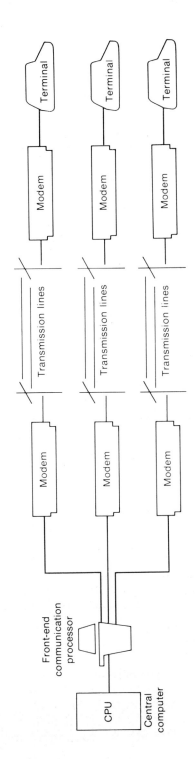

Figure 3-2:   Front-End/Central Computer Configuration

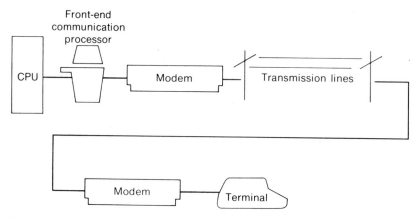

Figure 3-3:   Point-to-Point Network

units. These units are data handling devices that control the data transmission between a central computer and remote terminal sites. In theory, these hard-wired communication control units perform as many communication-related functions as possible in order to relieve the load on the central computer. Hardware manufacturers are moving away from nonprogrammable units, which are not flexible enough to meet current needs; therefore the emphasis in this book will be on configurations using programmable front-end communication processors. The reader should note that not all the functions listed below as performed by the programmable front-end communication processor can be performed by nonprogrammable units. This is why the nonprogrammable concept is being discarded in favor of the programmable concept, even though nonprogrammable devices often have a speed advantage over programmable devices.

Programmable front-end communication processors have become sophisticated, computer-like devices. In fact, many of these processors are full-scale minicomputers. They can be purchased with a variety of functional and logical options built in. Programmable front-end communication processors are many times more powerful than a hard-wired communication control unit, and have a wide range of capabilities. For example, if the front-end device is a general purpose minicomputer, then it can generally perform applications-type editing of input (editing for correct account number, field verification, and the like). On the other hand, if the front-end communication processor is programmable, but only designed for communications, then it probably cannot perform this type of applications program editing. Figure 3-4

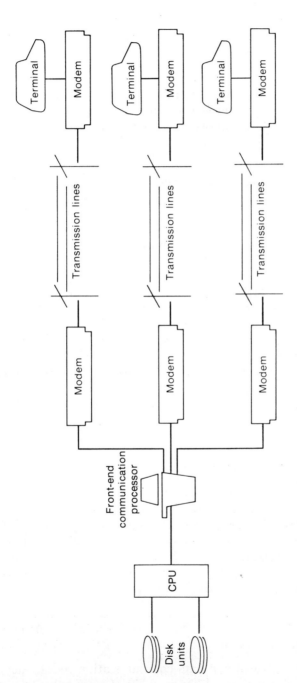

Figure 3-4: Network Using a Front-End Processor

shows a front-end communication processor controlling various input lines from remote terminal sites. The range of characteristics and functions of a programmable front-end communication processor is as follows:

- Connects from one to several hundred communication lines to the main computer. In this way it is the interface between the communication lines and the central computer. For very large networks there may be several front-end communication processors connected to one central computer.

- Accepts synchronous, asynchronous, or isochronous, serial or parallel transmission for data coming in from the remote terminal sites. In doing this, the front end must be able to convert whatever form of transmission is received into the form that the central computer is architecturally designed to accept. Conversion from serial to parallel character format can be performed by the central computer or the front end. This conversion from parallel format to serial format is known as serializing and deserializing the bits that represent each character. Most modern computers are word- or character-oriented machines. In other words, within the computer an entire character (6 to 9 bits) or a word of 2 to 4 characters is transmitted on each computer cycle. In data communications, an entire character is not transmitted all at once down the line (except in rarely used parallel transmission). Instead, the front end serializes the characters that it has received from the host computer. Serializing means disassembling a character or word and sending it down the transmission line one bit at a time.

- Polls terminals to inquire if they have a message to send or are in a state to receive a message. Polling is performed by the front-end device in order to relieve the central computer of the corresponding overhead work required to poll. Also, programmable front-end devices can be instructed to change the polling sequence whenever there are traffic peaks during the business day.

- Handles automatic answering or outward calling on the public switched network in order to connect various terminals to the system.

- Utilizes a variety of data communication codes whenever a network has many different types of terminals connected to it. The front end can perform the required code translation, although to be efficient this probably should be done by a specially built hardware device in order to reduce the overhead of the front-end communication processor.

- Performs circuit or line switching and provides a "store and forward" capability. Whenever one terminal is transmitting a message to another terminal that is busy, the front end may store that message and later forward it to the second terminal when it is not busy. Similarly, messages to a terminal that is out of order will be held until that terminal comes back on line.

- Handles transmission speed differences. Terminals that are connected to the system do not all have to transmit to the central computer at the same speed.

- Performs logging of all inbound and outbound messages on a special log. This provides data protection and accountability should a message be lost, and provides the capability to restart the system should a "crash" occur.

- Provides error detection and correction. The processor checks the accuracy of the received data, using the various techniques discussed in Chapter 5. When a message is found to be in error, the front end corrects or orders the retransmission of the message from the sending terminal. Also, the front end can intercept messages that have logical errors such as nonexistent terminal addresses or incorrect header information. The sending station is advised, or the message is routed to a central correction terminal.

- Adds communication line control codes to outbound messages and deletes these codes from incoming messages. Examples of such codes are end of block character, end of transmission character, start of message character, and the like.

- Addresses either a special group of terminals (group addresses), several terminals at a time (multiple address), one terminal at a time (single address), or sends a broadcast message simultaneously to all terminals in the system.

- Assigns serial numbers as well as time and date stamps to all messages that it handles.

- Is responsible for the "time-out" facility. If a specific terminal or station does not respond promptly, the system will "time-out" (skip over) that station and go on to the next station or activity that is to be performed so other users can be served even when one user is not responding correctly.

- Performs two levels of editing. In the first level, the front end may add items to a message, reroute the message, or rearrange data for further transmission. It may also check a message address for accuracy and perform parity checks. In the second level of editing, the front end is programmed to perform specific edits of the different transactions that enter the system. This editing deals with message content rather than form and is specific to the application programs being executed.

- Handles the message priority system, if one exists. A priority scheme is set up to permit a higher line utilization to certain areas of the network or to ensure that certain transactions are handled before other transactions of lesser importance.

- Performs correlations of traffic density and circuit availability. These analyses are mandatory for the effective management of a large data communication network. Some of the items included in a traffic density report are (1) the number of messages handled per hour or per day on each link of the network, (2) the number of errors encountered per hour or per day, (3) the number of errors encountered per program or per program module, (4) the terminals or stations that appear to have a higher than average error record.

- Determines alternative paths over which the data will be transmitted, if the network has multiple circuit links that can be chosen to transmit a message from one station to another. An alternative path may be chosen if the front-end senses an excessive line error rate or heavy traffic on one link.

- Performs miscellaneous functions, such as (1) triggering remote alarms if certain parameters are exceeded; (2) performing the multiplexing operations involved when circuits are utilized in a multiplexed format (this is usually done by specific multiplexing devices); (3) signaling abnormal occurrences to the central computer; and (4) slowing up the input and output of messages when the central computer is overburdened and there is a possibility of a systems crash due to heavy traffic.

In data communication systems, the central computer must be able to cope with a variety of situations. Therefore, many of these functions are transferred to the front-end communication processor in order to make the data communications economical, reliable, and easy to use. Multiple front-end computers may be used, each assigned to a particular type of communication or functional class of message. In this way, the central computer is left free to handle the actual functional processing of the various messages.

Port-Sharing Device

All front-end communication processors have a maximum capacity of circuits. The receptacles of the communication processor into which lines are connected are called ports. Therefore, if a front-end communication processor is designed to handle up to 100 ports, that means up to 100 circuits can be interconnected with it. Whenever a user wants to exceed the designed capacity of a front-end communication processor, a port-sharing device may be utilized.

A port-sharing device, or "bridge", treats several point-to-point lines as if they were a single multipoint line. In this way, it conserves ports at the expense of allowing only one terminal to transmit to the computer at a time. Figure 3-5 shows a port-sharing device with a local terminal (no modem required) and a remote terminal connected to it. Port-sharing devices are available that allow connection of up to four low speed terminals (either local or remote). In this way, one front-end communication processor port is expanded to handle four terminals. This type of device can be utilized to avoid installing a second front-end communication processor when all ports of the first are occupied. This may not be a long-term solution but it can be a short-term "holding action" until a new network can be configured or new hardware can be purchased. Port-sharing devices are digital bridges because they are installed on the digital side of the modem. Analog bridges, on the other hand, are used on the analog side of the modem whenever it is desired to make two or more point-to-point lines behave as a multipoint line. In actual practice, the common carrier bridges point-to-point lines together whenever a customer orders a multipoint line.

Line Splitter

A line splitter is similar to a port-sharing device except that it is located at the remote end of a line (see Figure 3-6). It is a "switch" that treats several terminals as if they were a multipoint line. The four terminals of Figure 3-6 share the total capacity (bits per second transmission rate) of the link, e.g. the line splitter shares the link between the four terminals.

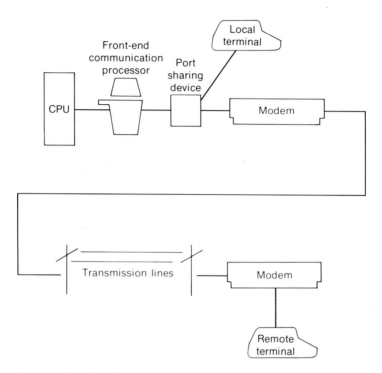

Figure 3-5:   Port-Sharing Device with a Local Hard-
Wired Terminal and a Remote Terminal

Modems

Modem is an acronym for modulator/demodulator (see discussion
in Chapter 2, under "Modulation"). A modem takes binary pulses
received from a computer, terminal, or other business machine
and converts those signals into a continuous analog signal that is
acceptable for transmission over a communication transmission
line. One such common method of transmission is called frequency shift keying (FSK) because a change in the binary value
transmitted is signaled ("keyed") by a change in frequency. The
0's and 1's from the computer are converted to tone signals of two
different frequencies. For example, assume the carrier signal on
the telephone line to be a continuous 1,700-Hz tone. Then, if a 1 is
to be transmitted, the 1,700-Hz carrier is shifted upward to 2,200
Hz; if an 0 is to be sent, the carrier is shifted downward to 1,200
Hz. The modem at the receiving end detects the frequency shift

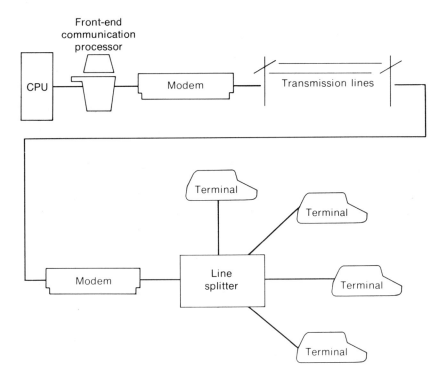

Figure 3-6: Line Splitter

and outputs a 1 or 0 as a consequence. The modems already shown in Figures 3-3, 3-4, and 3-5 all perform this type of function. Modems do not originate signals, rather they transform signals from computer binary form to analog form and back again.

Communication circuits are rated by the speed at which data can be transmitted over them. For example, voice grade telephone circuits are usually utilized at a set of fixed transmission rates ranging from 1,200 bps to 9,600 bps. In fact, the speed at which data are transmitted over these circuits is determined by the modem modulation rate. To transmit at 1,200 bps, the modem must be able to convert binary signals to analog signals at a rate of 1,200 bps. To transmit at 9,600 bps, most modems utilize special encoding schemes such as dibits (discussed in Chapter 2).

A modem can possess other features besides the modulation and demodulation of the data. Modems equipped with a special auxiliary unit can perform automatic dialing to call a remote terminal. They can be configured to be in a continuous ready state so they can be called from a remote terminal at any time; this feature is called "automatic answer." Certain modems can be used either for data transmission or alternate voice transmission. Some modems also allow simultaneous voice transmission. This is useful for troubleshooting between a central computer and a remote terminal site or in trying to synchronize various operations between a central computer and a remote job entry (RJE) station. Some modems also have a "reverse channel" capability, whereby a limited full duplex transmission capability can be achieved using two-wire circuits. In this configuration, while the modem is transmitting data in one direction (through the use of frequency multiplexing), the response character that acknowledges the receipt of an error-free message is simultaneously sent in the opposite direction over the same transmission line. This avoids turnaround time for the transmission of responses by the receiving station.

Modems are classified as low speed or high speed. Modems that operate up to 1,800 bps are usually rated as low-speed modems. These modems primarily utilize FSK. Modems in the range from 1,800 to 9,600 bps and beyond are generally termed high speed. These modems tend to use a type of phase modulation and a transmission methodology that employs dibits. The higher speed modems are usually used for remote CRT (video) terminals and RJE stations with multiple input/output devices (teleprinter, CRT, card reader, and line printer). The higher the speed of the modem, the more susceptible it may be to error, although phase modulation error rates are less than with FSK.

A unique type of modem (seldom used) is a parallel transmission modem. This modem has eight channels, assuming an 8-bit code is being transmitted, so that it can send 8 bits simultaneously. This type of parallel modem transmits a character at a time instead of a bit at a time and is utilized for CPU to CPU data communications. If the transmission speed of a regular modem were 1,200 bps and this arrangement was changed to a parallel modem, then the speed would be increased to 1,200 characters per second. This would be a substantial increase in the amount of data that could be transmitted between the two CPUs. Parallel transmission is not utilized over a long distance because the bits tend to drift forward and backward relative to one another and may interfere with each other, obliterating some of the bits from the previous character or the character following.

One special type of modem is the acoustic coupler. This is a popular type of modem that utilizes FSK. It is used often on the dial-up telephone network because of its low cost and convenience. Instead of coupling to the phone line electrically as other modems do, it couples acoustically. That is, the digital signals are converted to acoustic tones which are played into the mouthpiece of an ordinary phone. In the reverse direction, a microphone picks up tones from the earpiece and converts them to digital form. Figure 3-7 shows an acoustic coupler.

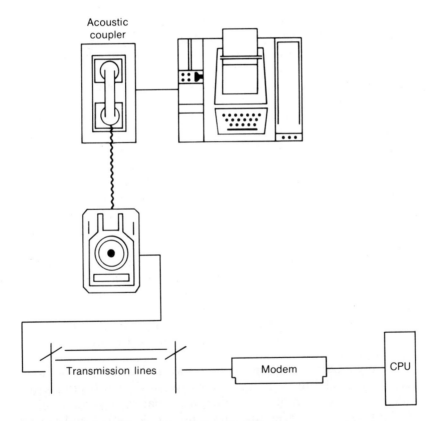

Figure 3-7: Acoustic Coupler at Remote Site

The major criteria for choosing a modem are transmission rate, turnaround time, error susceptibility, reliability, cost, and maintainability. The modem transmission rate must be sufficient to handle the basic system data volumes. Modem turnaround time is the length of time that is required for a modem transmitting in half duplex to shift from sending to receiving, or vice versa. The turnaround time may vary from 20 milliseconds to 200 milliseconds. This can become a large fraction of system time if many short messages are sent in alternating directions over a line. For example, consider that each time a terminal is polled there is a turnaround for the terminal to respond. Also, when using synchronous transmission, there is the "clear-to-send delay," which is the time required to synchronize the two modems prior to transmitting any data bits. The response time, especially on polled lines, can be degraded by lengthy modem turnaround times.

Error rates on modems depend basically on speed and the type of modulation utilized. Phase modulation is less error prone than frequency modulation. When the speed of a modem is in the range of 4,800 to 9,600 bps, older modems require conditioned voice grade lines to reduce the errors during transmission. Conditioning is the electronic balancing of line characteristics by the common carrier and is supplied at extra cost. Some modems dynamically balance the line to reduce the error rates, although this, too, is an added cost feature in the modem.

The cost of a modem is directly proportional to the speed at which the modem transmits. Acoustic couplers are inexpensive but they usually transmit in the 110- to 1,800-bps range. Reliability and maintainability are fundamentally important to the success of a data communication network. First, the modem must be reliable. Most manufacturers will provide information about the history of mean time between failures (MTBF). Availability of maintenance service in the locality where the modem is installed is also an important factor. Some modems offer an alternative lower transmitting speed to help overcome errors produced by a line that has become noisy. Another similar option for private-line modems is the capability to transmit over the dial network if the private line fails (Figure 3-8).

Many modems offer special maintainability features to facilitate "loop-back" tests. A loop-back test involves reflecting back and measuring a signal (either digital or analog) at various points in the network in order to test for proper operation and isolate failures. Figure 3-9 shows several points in a network at which a loop-back test may be performed with the aid of such modem facilities. Note that if loop-backs A, B, and C are all right, then the three modems are functioning correctly, but if loop-back D fails, then the problem is in the line connecting Modems I and II (this

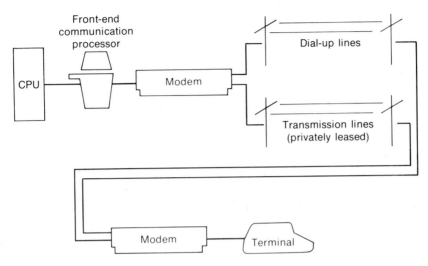

Figure 3-8: Dial Back-Up Configuration for Private Lines

assumes loop-back G was okay). Loop-backs A through G are analog; loop-backs W, X, Y, and Z are digital.

Multiplexing

Multiplexing is the term used to describe the combination of two or more signals within a single channel. Multiplexing is used primarily to save circuit costs and to use resources more efficiently. A typical multiplexed communication configuration is shown in Figure 3-10.

Some of the reasons for using multiplexing are that:

- There are inherent limitations on the capacity of any communication channel.
- The common carriers offer more or less fixed packaging of communication channels, which results in channels of varying capacity (e.g. 2,400, 4,800, etc. bps). In general, a "quantity discount" is available for channels of larger capacity.
- The demand for channel capacity often seems to come in different increments than do the capacity packages offered by the common carriers.
- The natural characteristics of communication demand which make multiplexing attractive. In most networks

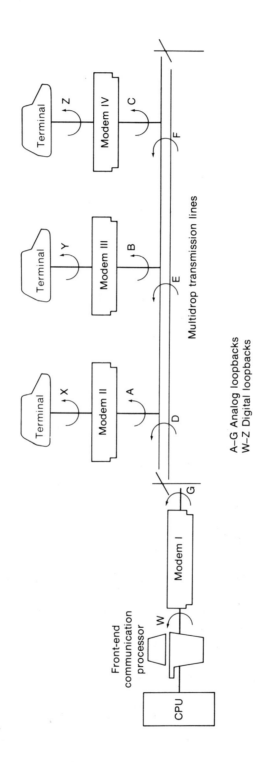

Figure 3-9:   Fault Isolation with Loop-Back Tests in a
Multidrop Line Configuration

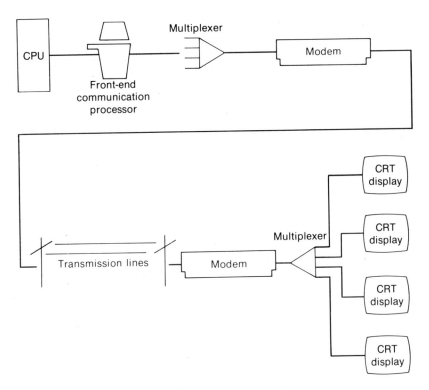

Figure 3-10:   Network Multiplexing

there is some natural "trunking," i.e., the simultaneous channeling of several messages between the same terminal points.

Types of Multiplexing

There are two basic types of multiplexing: frequency division multiplexing (FDM) and time division multiplexing (TDM). To understand the distinction between these types it is helpful to use the concept of a channel having two dimensions, frequency and time. For example, Figures 3-11, 3-12, and 3-13 present different types of signals carried by a channel and depicted as a frequency versus time graph. Figure 3-11 shows a continuous tone of frequency f_1. It is represented by a line parallel to the time axis at the value f_1. Figure 3-12 shows an intermittent tone of the same frequency. Again, the line is parallel to the time axis at a value of

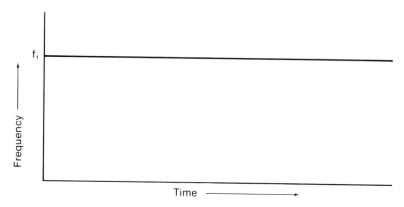

Figure 3-11:   Continuous Tone

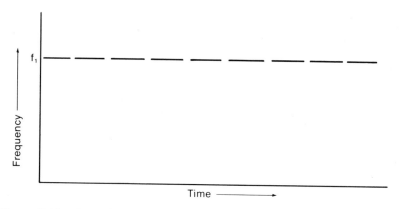

Figure 3-12:   Intermittent Tone

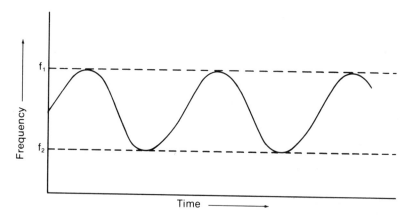

Figure 3-13:   Warbling Tone

f_1, but the line is discontinuous, representing the intermittent (on/off) nature of the tone.

Figure 3-13 shows a "warbling" sound such as might be produced by an emergency vehicle with an electronic siren. In this figure, the excursions between the frequencies f_1 and f_2 are represented by the oscillating line. This form of depiction will be used in the following discussion.

Each type of channel has a characteristic limitation in bandwidth, i.e., no channel can pass an arbitrarily wide range of frequency. The human ear, for example, is a communication device that cannot respond to frequencies in excess of about 18,000 to 20,000 Hz. (This is why "silent" dog whistles are silent.) Conversations, however, do not require a bandwidth of more than 3,000 Hz and human speech can be quite intelligible even if the bandwidth is restricted to 1,000 or 1,500 Hz. Thus, much of the capacity of the human ear is "wasted" in processing conversation because the capacity above 3,000 Hz is not even required. Nothing can be done about this waste for acoustical communications. However, techniques can be used to make use of the similar capacity in electrical communication channels.

Frequency Division Multiplexing (FDM)

Frequency division multiplexing is the older of the two multiplexing technologies. It has been used for over 30 years. FDM "stacks up" the signals in the frequency domain. To understand how frequency division multiplexing works, use the analogy of how a voice grade phone line of 3,000 Hz usable bandwidth might be divided into four separate communication channels using a frequency shift keying (FSK) technique. Frequency shift keying was discussed in the section covering modems.

Figure 3-14 depicts how a 3,000 Hz voice grade channel can be divided into four separate transmission paths (this is for analogy only and the frequencies utilized here do not depict those used in real life).

Assume that this voice grade line is divided into four transmission paths that have carrier waves of 600 Hz, 1,200 Hz, 1,800 Hz, and 2,400 Hz. Notice that the carrier wave utilized for the upper channel (2,400 Hz) can be modulated up to 2,600 Hz (represents a binary 1) or down to 2,200 Hz (represents a binary 0). Using this same logic you can see how the 1,800 Hz, the 1,200 Hz, and the 600 Hz channels could also be simultaneously transmitting intelligence, such as a binary 0 or a binary 1.

Reviewing this figure further, you can see that if there were four terminals (A, B, C, and D), these four terminals could be

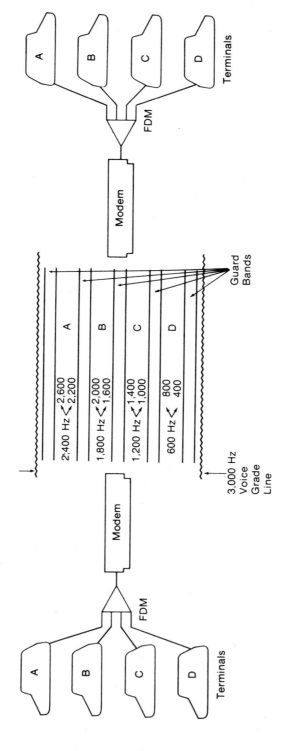

Figure 3-14:   Frequency Division Multiplexing (FDM)

multiplexed onto one voice grade line and at the receiving station the multiplexer would demultiplex the signal and pass the appropriate data to the appropriate terminal. In practice, to prevent interference between adjacent multiplexed signals, there would be narrow guard bands (in our analogy, 200 Hz wide).

In summary, this technique (FDM) splits the bandwidth into several smaller bandwidths at different frequencies. Each of these signals is assigned a discrete slot (narrow frequency range) within a particular bandwidth and then all signals are transmitted simultaneously.

Other characteristics of frequency division multiplexing are that the subchannels need not all terminate at the same location. FDM can be used on multidrop networks where you drop off each frequency at a station and then continue only the remaining frequencies to the more distant stations. Once you determine how many subchannels are required, it may be difficult to add more subchannels because all the frequencies in a group must be changed. For example, if you were to purchase or lease a 12-channel FDM group and two years later you needed to increase it to a 16-channel FDM group, you would have to get a completely new frequency division multiplexer because all 12 frequencies would have to be changed in order to accommodate the four new frequencies.

Time Division Multiplexing (TDM)

Time division multiplexing could be called time slicing or sharing the timing of the communication circuit with various terminals. In time division multiplexing, the multiplexer uses a high-speed data stream to interleave the bits or characters from several slower data streams. For example, if four terminals were transmitting at 300 bps each, then the time division multiplexed bit stream would have to be transmitting at 1,200 bps.

By looking at Figure 3-15, you will see that terminals A, B, C, and D will each transmit a character through the time division multiplexer; the multiplexer will assemble the characters in a frame, and send the frame down the transmission path. If each of the four terminals were transmitting at 10 cps, the transmission path would have to be capable of moving data at 40 cps (the modem would drive the data). Note how each of the characters go down the transmission path in a frame. After the frame is transmitted down the transmission path, the multiplexer at the receiving end demultiplexes this frame to give the proper character to the proper terminal.

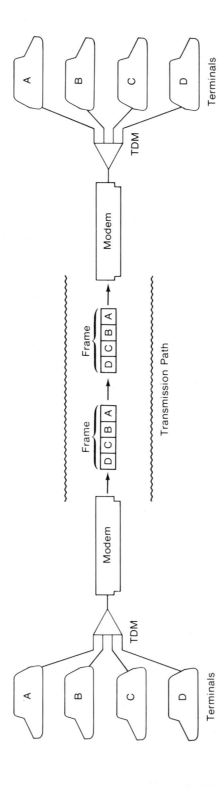

Figure 3-15:   Time Division Multiplexing

Some time division multiplexers use a fixed frame approach where each frame consists of one character or bit for each channel (character multiplexing is usually preferred because it is more efficient). Other time division multiplexers use a variable frame approach, where the data elements from each input terminal need not be present in each frame because control information is transmitted with each frame to indicate which channels are present.

Other characteristics of time division multiplexing are that it is generally more efficient than frequency division multiplexing because more subchannels can be derived. It is not uncommon to have time division multiplexers that will share the line with 32 different low-speed terminals. Also, it is easy to add channels to a time division multiplexer after you have established your base group. For example, if you were to purchase a 12-channel base group and two years later wanted to expand to a 16-channel base group, you would have to buy only 4 more channels and have them field installed to change the timing characteristics. Time division multiplexing channels usually all originate at one location and all terminate at another location. It is usually too costly to have a multipoint drop-off utilizing time division multiplexing.

Some multiplexers have a contingent feature where you can connect more terminals to the multiplexer than it is able to handle. This means that if the multiplexer is being used to its fullest capacity a terminal trying to establish a connection will be unsuccessful (it will get a busy signal).

Concentrators

Concentrators are used for the same purposes as multiplexers. They function by manipulating signals logically. A concentrator takes in messages from several sources and puts out fewer message streams that it takes in. The transformations performed by a concentrator may include:

- Increasing output transmission speed over input speed
- Coding changes and error checking
- Adding source identification data to messages
- Temporarily storing characters, words, or entire messages
- Deleting extraneous characters, such as SYNC characters
- Changing message formats
- The inverse of any of the above for message traffic bound in the opposition direction
- Polling and buffering messages.

Concentrators differ from multiplexers in two ways. First, a multiplexer is "transparent"; that is, people at the terminals cannot tell whether a multiplexer is installed on the line. Concentrators are not transparent; they permanently alter the message stream. Second, concentrators are "intelligent" in the sense that they examine message content, react to it, and store data. They are, in other words, more computer-like than multiplexers and some are, indeed, minicomputers. For a multiplexer, N input lines generate N outputs, whereas for a concentrator, N input lines generate M outputs where $M \leq N$. Thus concentrators may buffer messages for transmission at a later time.

Concentrators are generally applied where their "intelligence" is likely to pay off. A good example is in international communications where channel costs are very high. A concentrator (generally a minicomputer) is located at a "gateway" (terminus for an overseas line) and receives messages from several domestic channels. These channels are usually operating at far less than capacity, and thus the concentrator output is driven at or near full capacity by the average of the sum total of the input rates. The concentrator stores messages temporarily until it can find space for them on the output channel.

Concentrators tend to be used in situations that have more sophisticated needs than multiplexers can meet. Costs are usually higher and economic justification is a more complex and specialized process because concentrators may control portions of a larger network.

Categories of Terminals

During the design of a data communication network, one of the decisions to be made is the determination of the type of terminal best suited to the system. This is a decision with far-reaching consequences. The choice of terminal is of equal importance with the choice of the central computer, because it is at the terminal that the human interface to the system is consummated. As a system grows and new pieces of hardware are added or old hardware is upgraded, it is usually transparent to the terminal user. Only when the terminal itself or its supporting software is changed or upgraded is the terminal operator directly affected. For this reason, the choice of a terminal should be predicated on both the current requirements of the system and the expected future growth. As the system matures, the terminal operator becomes accustomed to the specific terminal and is naturally reluctant to change to another terminal. The system designer, therefore, should start with a terminal that will be adequate for many years to come while still meeting the current needs of the remote user.

There are five basic categories of data communication terminals:

- Teletypewriter terminals
- Video terminals
- Remote job entry terminals
- Transaction terminals
- Intelligent terminals.

Teletypewriter terminals are typewriter-like terminals that have keyboards and print hard copy. These terminals have no programmable capability and are used primarily on low-speed leased or dial-up lines. Teletypewriter terminals are usually unbuffered although some may have the capability to buffer one line of print to increase their printing speed. These terminals may print by the impact method or use heat-sensitive paper or ink-jet character forming. They print one character at a time at speeds that range from 10 characters per second to roughly 150 characters per second. This is the most common type of terminal, although in the future the video terminal may overtake this segment of the market. Teletypewriter terminals will continue to find widespread uses for certain applications, especially in dial-up situations where acoustic couplers are used.

Teletypewriter terminals generally have three types of keys: (1) regular text keys that include A-Z, 0-9, and the special characters; (2) control keys for the communication control codes such as START, STOP, DELETE, END OF TRANSMISSION, and the like; (3) function keys, which include the typewriter-like functions of carriage return, backspace, horizontal tab, and so on. Some of these terminals have punched paper tape readers and punches attached as auxiliary devices that are logically independent but functionally related.

The lease cost for teleprinter-type terminals ranges from approximately $75 per month upward. The currently popular 30 character-per-second devices cost approximately $100 per month to lease.

Video terminals, sometimes called Cathode Ray Tubes (CRT), are available in two main types, alphanumeric and graphic. The less common graphical video display terminals are capable of depicting line segments of various lengths and directions to facilitate the drawing of pictures by computers. For example, interactive graphic terminals can be used to develop chart plotting systems and design the contours of automobile body styles for new designs. Interactive graphic terminals may have a light pen or other graphic input device that the terminal operator uses to draw on the screen of the video display terminal.

Alphanumeric video display terminals are used primarily in business-oriented data communication networks. These terminals come in two configurations, "stand alone" and "cluster." Stand alones are individually self-sufficient. Cluster units share a common co-located controller to reduce costs. These terminals are usually buffered and provide some flexibility for operator or computer control of formats. They are buffered to the extent that the individual terminal itself or a video display control unit can hold one or more screens of data. Figure 3-16 shows a cluster configuration of five CRT displays and an auxiliary printer.

Video display terminals are usually utilized on private leased lines but they also have wide application on dial-up lines. These display terminals may also have a printer attached to make permanent copies of selected data. These printers usually are of the Teletypewriter terminal variety although a system that consists of a cluster of video displays may also have a high-speed line printer attached to the controller that can be called into use by any of the video display units.

Most video display terminals utilize a keyboard that is very similar to the standard typewriter keyboard, plus the appropriate control and function keys. Video terminal display speed is generally a function of line speed, and ranges from approximately 10 characters per second to 1,200 characters per second. Video terminals have a marker, usually an underline bar, that appears below the next character position into which a character will be inserted. This marker is called a cursor. There are usually special function keys that will move this cursor both vertically and horizontally so the terminal operator can locate it in any position that is desired. Some video terminals have a light pen to allow the operator to touch the screen in a manner analogous to moving the cursor to perform editing operations.

A "refresh" memory built into the stand-alone terminal or in the cluster-type control unit provides the buffer storage for one or more screens of data. Just like a home TV, video screens must be refreshed at least 30 times a second or the image will present an annoying flicker.

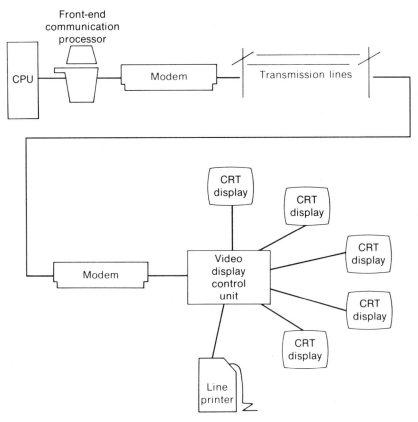

Figure 3-16:   Video Display Unit with a Cluster Configuration of Five CRT Terminals and One
Printer

It is possible to obtain minimally configured video display
terminals that emulate Teletypewriter terminals for about $75
per month if no hard copy output capability is required. A
minimally configured video terminal with its own attached 30-
character-per-second printer leases for approximately $175 per
month. More sophisticated video display devices with some programmable features, hard copy output, local editing, and the like
range upward from $175 per month.

Remote job entry (RJE) stations usually operate at 2,400, 4,800,
or 9,600 bps over voice grade lines because of the larger quantities of data that must be transmitted. Figure 3-17 shows an RJE

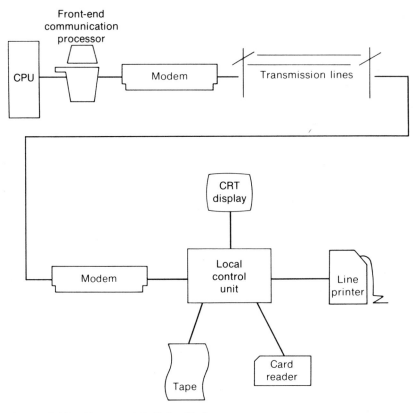

Figure 3-17:   Remote Job Entry Station

station consisting of a control unit, a card reader, a line printer, and operator console (CRT display), and possibly paper tape, magnetic tape, or disk storage capacity. RJE stations are almost always buffered, and are frequently programmable. Some of the more sophisticated RJE stations are basically minicomputers, so they can perform some of the local processing that the business requires and only pass on to the central computer the processing that they cannot handle.

Prices for RJE stations currently vary over a wide range, starting at a minimum of $200 to $300 per month, for a limited configuration, and working up to $700 to $1,000 per month for a configuration that has full capabilities of reading and printing in a buffered, error-controlled system. Many minicomputer-based RJE systems are even more expensive because they include disk storage and magnetic tape capabilities.

Transaction terminals are low-cost terminals that are driven by buffered, shared controllers or a minicomputer that is located within the transaction environment, generally a retail store. Transaction terminals usually use privately leased lines or hard-wired local loops. Figure 3-18 depicts transaction terminals that are on a local loop located within a retail store. The most common transaction terminals are the credit authorization terminals that are utilized by many retailers. Another, more sophisticated, type of transaction terminal is the "point-of-sale" terminal. Such terminals have the special capability of reading bar codes printed on the product being sold, and of ringing up the sale in such a way that pricing and updating of the store's inventory is accomplished. These terminals are buffered and are designed around a particular industry application such as retail point of sale, banking terminals, credit checking, or supermarket checkout. These transaction terminals range from $20 a month (credit checking terminals) to several hundred dollars a month for a point-of-sale cash register and its accompanying equipment to read product codes.

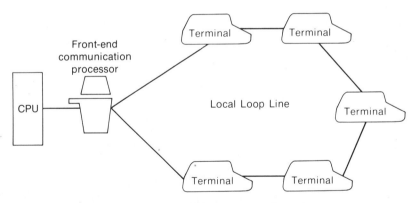

Figure 3-18: Transaction Terminals in a High-Speed Loop

"Intelligent" is a term applied to terminals (often video terminals) with a small, built-in computer. They are significantly different from the terminal categories discussed previously because they provide substantial amounts of useful functional capability without any connections to other higher level devices. In fact, some intelligent terminals challenge minicomputers in

the amount of their capability. The differences between intelligent terminals., minicomputers, and RJE stations will probably almost disappear in the next few years.

Intelligent terminals usually have buffers, are programmable, and are modular so that components such as disks and tapes can be added easily. They have substantial functional capability that is independent of the central processing unit. In other words, they can be used for local data entry but can perform transaction editing, verification of data fields, some actual processing or data base inquiry to their own local data base, and pass on certain types of processing to a central computer. Intelligent terminals can function as a combination of RJE stations, communication concentrators, application processors, and intelligent cluster control units. Figure 3-19 depicts an intelligent terminal that is

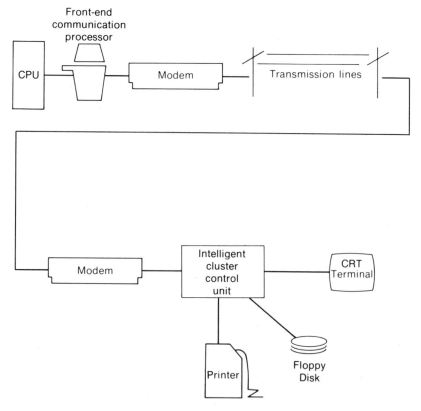

Figure 3-19:   Intelligent Terminal as an Intelligent Cluster Control Unit

utilized as an intelligent cluster control unit. In this application, it could control Teletypewriters, video terminals, transaction terminals, tapes, disks, and even RJE stations.

Because of their programmability, intelligent terminals are usually the most expensive form of terminal. The cost of an intelligent terminal ranges upward from $200 or $300 per month until it merges with RJE stations and minicomputers.

Figure 3-20 offers a summary of the five categories of data communication terminals, and Figure 3-21 is a checklist of terminal characteristics. Probably, only intelligent terminals would possess all the characteristics listed in Figure 3-21, but the systems designer can use it as a checklist, regardless of the type of terminal being considered for the data communication network.

- **Low-Speed Teletypewriters (0-1200 bps)**
  - Buffered or unbuffered
  - Limited intelligence
  - Used on dial-up or leased lines
  - Popular applications — Time sharing, message switching

- **Low-, Medium-, and High-Speed Video Displays (300-9600 bps)**
  - Alphanumeric or graphic
  - Mostly buffered with moderate intelligence
  - Used on dial-up or leased lines
  - Popular applications — Fast-response data base inquiry systems

- **Remote Job Entry Systems (2400-9600 bps)**
  - Card reader, printer, operator console as a minimal configuration
  - CRT display, tape, diskette capability optional
  - Mostly buffered, frequently programmable
  - Used on dial-up or leased lines
  - Popular applications — Card reader, line printer, access to batch job queue

- **Transaction Terminals (300 bps-50,000 bps Loop Line)**
  - Low cost per workstation, driven by buffered, shared controllers
  - Mostly buffered and designed around particular application
  - Used mostly on leased lines
  - Popular applications — Retail point-of-sale, banking, credit checking, supermarket checkout

- **Intelligent Terminals (2400-9600 bps)**
  - Buffered, programmable, highly modular
  - Substantial functional capability independent of host CPU (e.g., local data entry, transaction edit and/or verification, and data base look-up independent of host CPU)
  - Can function as combination of remote data entry station controller, remote display controller, communications concentrator or applications processor
  - Can control such devices as teleprinters, CRTs (both local and remote), transaction terminals, tapes, diskettes, disks, and on-line storage
  - Used on either dial-up or leased lines and can perform substantial functions without connection to host CPU

Figure 3-20:   Categories of Teleprocessing Terminals

- Buffered/unbuffered
- Send only, receive only, send/receive
- Human factor considerations
- Operating speeds (characters per minute)
- Hard/soft copy
- Portability (mobility)
- Modes of operation (HDX/FDX, lease/dial circuits)
- Code format capabilities
- Error control features
- Security features
- Peripheral attachments possible
- Unattended operation
- Off-line capabilities
- Message formatting capabilities

Figure 3-21:   Checklist of Terminal Characteristics

This checklist, which is a summary of the characteristics discussed previously, should be used to ensure that the designer considers each terminal characteristic for the application at hand.

Hardware Interfaces

After a rather chaotic start in the late 1950s and early 1960s, hardware manufacturers have developed standards for interfacing equipment. The most important of these for U.S. data communication network designers is the RS-232 interface. It specifies the physical and logical standard for interfacing communication devices such as terminals, and modems. This industry standard ensures that a modem purchased from one manufacturer will be able to function with a terminal purchased from another manufacturer. The designer is well advised to use this standard interface, almost without exception. There may be cases where it should not be used, but those cases should be carefully researched and justified.

Terminal Security

Terminals are designed to give easy access to the computer by the user. In some instances, they serve that purpose too well by providing an opportunity for dishonest or incompetent users to access a system, and especially its data base. The designer must attend, therefore, not only to the functional and service-life characteristics of terminals when making design decisions, but also to the security aspects. The security of computer-communication systems is a far-reaching subject, beyond the scope of this book, but a few elementary considerations can be mentioned:

- Can physical access to the terminal be limited? (Locking on-off switch, etc.)
- Can logical access to the computer be limited? (Nonprinting passwords, unique terminal identifications, security codes, etc.)
- Can access to the communication link be limited? (Cryptography).

Consideration of these questions in the light of system requirements is essential to ensuring that the proper terminal is chosen for the application. Figure 3-22 is a checklist of security considerations for on-line terminals.

- Unique terminal identification by the computer
- Lockable keyboard
- Nonprinting feature when keying security code
- Identification card reader
- Signature verification
- Cryptography
- Physical lock on terminal on/off switch
- Physically secure location for terminals
- Communication facilities that may be locked in an inoperative state
- Transaction code terminals
- Departmental user number
- Individual security code
- User not authorized for a specific program, file, or record

Figure 3-22:   Security Considerations for Terminals

Questions—Chapter 3

The hardware building blocks of data communication systems should now be familiar. Some terms that, to the reader, were previously "buzz words" now have some meaning and reality. There are four kinds of questions: True or False, Fill-In, Multiple Choice, and Short Answer. Of the short answer questions, the first five test your understanding of the major characteristics of the hardware building blocks. When you develop a confident, insightful grasp of these characteristics, you will be able to fit the blocks together in effective, and perhaps innovative ways. The first five short answer questions are open book; the last four are "fact feedback" questions. Try them first as closed book questions.

True or False

1. General purpose computers are generally designed with built-in communication interface hardware.

2. A communication controller is a programmable front-end processor.

3. Nonprogrammable devices often have a speed advantage over programmable devices.

4. Nonprogrammable devices can perform all the functions programmable devices can perform.

5. A bridge treats several point-to-point lines as if they were a single multipoint line.

6. Modems are used to originate signals.

7. Besides converting signals from binary form to a continuous analog form, the other primary function of a modem is serialization and deserialization of data bits.

8. A modem that operates between 1,800 to 9,600 bps would usually be rated as a low speed modem.

9. TDM may only be applied to FSK signals.

10. Teletype terminals generally have three types of keys.

Fill-in

1. A front-end module can take ——— forms.

2. A ——— ——— computer configuration is designed to handle a specific set of communication facilities and terminals.

3. The ——— function is being performed when the front-end calls terminals to inquire if they have a message to send or are in a state to receive a message.

4. If a specific terminal or station does not respond promptly, the system will ——— ——— or skip over that

station and go on to the next station or activity that is to be performed.

5. The receptacles of the communication processors into which lines are connected are called ——— .

6. Modem is an acronym for ———/——— .

7. FDM stands for ——— ——— ——— .

8. It is at the ——— that the human interface to the system is consummated.

9. ——— is the term used to describe the combination of two or more signals within a single channel.

10. The type of modem that is not connected electrically to a phone line is called an ——— ——— and is often used because of its low cost and convenience.

Multiple Choice

1. In which of the following computer figurations is a general purpose computer least likely to be used for both batch and timeshared processing.
   a) Front-end configuration
   b) General purpose computer configuration
   c) Stand-alone computer configuration
   d) A and B above with equal probability
   e) B and C above with equal probability

2. Correlation of traffic density and circuit availability would include all the following EXCEPT
   a) Number of errors encountered per hour or per day
   b) Number of errors encountered per program or per program module
   c) Number of messages handled per hour or per day on each link
   d) Which terminals or stations appear to have a higher than average error record
   e) No exceptions—all the above would be included in correlation of traffic density and circuit availability.

3. The feature of facilitating loop-back tests would primarily be included in a modem to:
   a) Increase transmission rates
   b) Decrease turnaround time
   c) Decrease the original cost of the modem

    d) Ease the isolation of errors in the network

    e) All the above

4. All the following characteristics apply to concentrators EXCEPT:

    a) Costs are usually higher than in situations where multiplexers could be used

    b) If there are N input lines and M output lines then M is less than or equal to N.

    c) They are intelligent

    d) They are transparent

    e) All the above characteristics would apply to concentrators.

5. The designer of a data communication system should consider which of the following in choosing a terminal to be used for the system?

    a) Functional characteristics

    b) Security aspects

    c) Service life

    d) All the above

    e) None of the above

6. Which of the following types of terminal has the highest bits per second capacity?

    a) Intelligent terminal

    b) Remote job entry system

    c) Teletypewriter

    d) Transaction terminal

    e) Video display

7. Which type of terminal is always programmable?

    a) Intelligent terminal

    b) Remote job entry system

    c) Teletypewriter

    d) Transaction terminal

    e) Video display

8. The type of configuration employed in situations primarily where the input/output and computing processing requirements are very large and where rapid response time is of the essence is:

    a) Front-end configuration

    b) General purpose computer configuration

c) Stand-alone computer configuration
d) A and B above with roughly equal probability
e) B and C above with roughly equal probability

Short Answer

1. Find ads in data processing or data communication journals that depict data on a Teletypewriter terminal, a video terminal, and a transaction terminal. Make a table, listing major terminal characteristics (you select which ones) as rows and the three terminal types as columns. Fill in the row-column intersections with the appropriate data. If you can, state some conclusions drawn from the comparisons shown.

2. List, briefly illustrate, and discuss some of the characteristics that make a computer good for "number crunching," and for communication. Draw some conclusions if possible.

3. List some systems applications in which an automatic answer capability is necessary; similarly, list some systems applications in which an automatic calling capability is necessary.

4. Three terminals (T_1, T_2, T_3) are to be connected to three computers (C_1, C_2, C_3) so that T_1 is connected to C_1, T_2 to C_2, and T_3 to C_3. All are in different cities. T_1 and C_1 are 1,500 miles apart, as are T_2, C_2, and T_3, C_3. The points T_1, T_2, and T_3 are 25 miles apart and the points C_1, C_2, and C_3 are also 25 miles apart.

$$
\begin{array}{lll}
25\text{ miles}\left\{\begin{array}{l} T_1 \text{ ——— } 1{,}500\text{ miles ——— } C_1 \\ T_2 \text{ ——— } 1{,}500\text{ miles ——— } C_2 \end{array}\right. & 25\text{ miles} \\
25\text{ miles}\left\{\begin{array}{l} T_3 \text{ ——— } 1{,}500\text{ miles ——— } C_3 \end{array}\right. & 25\text{ miles}
\end{array}
$$

If telephone lines cost $1.00 per mile, what is the line cost for three independent lines? If a multiplexer/demultiplexer pair costs $2,000, can you save money by another arrangement of the lines? If so, how much?

5. What is the difference between a front-end communication processor and a modem?

6. What is the term used to describe the combination of placing two or more signals on a single channel?

7.   For one piece of hardware mentioned in this chapter, N input lines generate M outputs where M is less than or equal to N. Name this piece of hardware.

8.   What are the five categories of terminals?

9.   What is the designation of the standard interface for the connecting of data communication equipment?

4

Network Configuration Concepts and Control Techniques

Having once acquired the basic facts about methods
and equipment for data communications, the designer
must learn how that equipment interconnects with
networks. The concepts and control techniques utilized
in the design of data communication networks depend
upon both the network functional requirements and the
hardware employed. Appropriate configurations and
controls can reduce costs and improve network
efficiency and throughput. When the network designer
begins to develop alternatives he must analyze the
various configurations and control techniques in order
to achieve a cost effective network.

Network Configuration

Networks can be classified as "switched" or "non-switched."
Information routings change with time in a switched network;
they remain constant in a non-switched network. Switched networks will be discussed in a later section of this chapter. Non-
switched networks are built of two kinds of building blocks;
point-to-point lines and multipoint ("multidrop") lines.

Multidrop Lines

Point-to-point lines are just that; they have only two end points usually occupied by a computer and some kind of terminal, respectively. To reduce the cost of the communication network, it is often desirable to attach more than one terminal to a single communication line. A line with several "drop points" is called a multidrop line. Figure 4-1 depicts a multidrop configuration.

On multidrop lines, only one terminal can transmit at any one time unless multiplexing is utilized. Each terminal must have an address and it must have the ability to recognize that a message is being sent to that address. For example, a line may have three terminals with addresses A, B, and C, as in Figure 4-1. The central computer may send a message down the line, addressed to Terminal B. (The address precedes the message.) Each terminal has circuitry that "listens" on the line for its own address; Terminal B will recognize its address and the other terminals will ignore it. Therefore only Terminal B will receive the message.

Multidrop configurations also usually have the capability to make use of group or broadcast addresses, in which more than one terminal on a multidrop line can receive a message. In group addresses, a message is transmitted containing the addresses of a preselected group of terminals on a multidrop line. In broadcast addresses, a message is sent down a multidrop line with an address that can be recognized by all terminals sharing that line.

Another form of multidrop line configuration is shown in Figure 4-2, Loop Lines. This type of configuration uses a short (approximately 1000 to 2000 feet) loop line; therefore no modems are required. In this multidrop configuration, a data stream is transmitted around the entire loop at speeds of approximately 50,000 bits per second. Messages are preceded by the address of the terminal to which they are directed. These messages circle through the loop and, as a terminal recognizes its own address, it deletes its message from the loop and processes it. Local loop lines are very effective in situations where several terminals are located close together. Such schemes are currently being utilized in factory production control, retail point of sale, and banking. This form of multidrop line utilizes synchronous transmission and requires more complex terminal logic to recognize its address.

Network Line Control/Polling

Network line control includes the operating procedures and non-text signals by which message transmission over a data communication network is controlled. In other words, line control

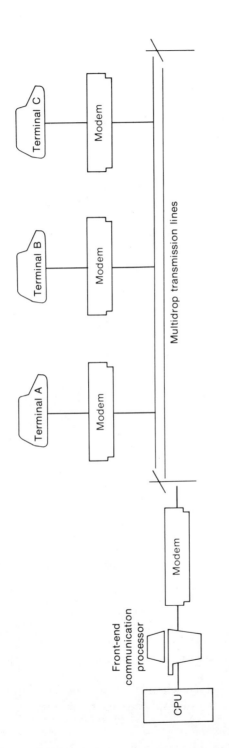

Figure 4-1: Multidrop Configuration

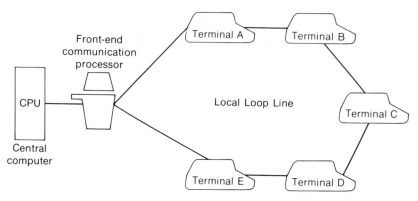

Figure 4-2: Loop Line Configuration with Terminals
             Located Short Distances from One Another

deals with allocating a limited resource (network capacity) in ways that will ensure that users receive allocations that meet their needs. There are two basic approaches to network line control: contention, and central control. A contention system is a data communication network in which terminals seize network facilities on a "first-come, first-served" basis. The most familiar example of a contention system is a dial telephone. If a call is in progress, no other call may be received. If two calls are made to a phone at roughly the same time, the one that first seizes the line will get through. Therefore, contention is a term that is applicable if several terminals share a common line or some other single resource, e.g., a program in a computer. One terminal makes a request to transmit data and if the line in question is not in use, the transmission proceeds. If the line is busy, the terminal must give up, and try again.

In a central control system, a single entity, usually the central computer, controls all message transmissions both to and from all terminals. The central computer withholds from the terminals the ability to independently gain access to transmission lines. The central computer gives the opportunity to transmit to one terminal at a time or to a remote concentrator that, in turn, gives the opportunity to transmit to one terminal at a time for the several terminals that may be connected to it. Most large data communication networks using private lines operate on the central control philosophy. This process is called "polling."

Polling takes place when the front-end communications processor, or a concentrator site calls each terminal in its domain to see if there are any messages to be transmitted. If the terminal being called has a message to transmit, the front-end processor, or concentrator relinquishes control of the line to that terminal. When the message has been transmitted, control is then resumed by the front-end processor, or concentrator. This technique of line control ensures that all terminals in a network have an opportunity to either transmit messages to or receive messages from the central computer. Even though the user at the remote terminal thinks that he is transmitting a message when he depresses the transmit key, he is not, because the message is not really transmitted until the central system polls the terminal in question and relinquishes control to that terminal. The system must be able to poll each terminal often enough to ensure that all messages can be gathered in without excessive delays. The scenario of a polling scheme follows this pattern: The central computer sends a message saying, "Terminal A, do you have anything to transmit? If so, send it." If Terminal A has nothing to send (i.e. a negative reply received by the central computer) the next polling message will be sent. "Terminal B, do you have anything to transmit? If so, send it." And so on until all the terminals in the network have been polled.

The polling device maintains in its memory a polling list giving the sequence in which the terminals should be polled. Some nonprogrammable front-end communication control units have hardware circuits that perform polling. These hardware circuits are sometimes modifiable through the use of pluggable control wires. This polling list is used to determine the sequence and priority with which terminals are polled. High volume terminals may have their addresses listed more than once on the polling list and thus be polled more frequently than the other terminals.

The most common form of polling is "roll call" polling, in which the polling list is used to "call the roll" of the network terminals. If there are 20 terminals in a network and they are polled sequentially from 1 through 20, each terminal will have equal priority to transmit and receive messages. In order to give a higher priority to some terminals, this sequential roll call polling list might be modified to include the address of Terminals 10 and 15 more often. For example, the roll call polling list might be developed as follows: 1, 10, 2, 15, 3, 10, 4, 15, 5, 10, 6, 15, and so on.

Another form of polling is "hub-go-ahead" polling, in which the polling device addresses the terminal at one end of the line and each terminal passes the polling message down the line. Figure 4-3 shows the central computer passing the poll down to the end of the line (Terminal A) and that poll being passed successively back to the central computer. Hub-go-ahead polling reduces the number of line turnarounds necessary on a multidrop line by making each terminal relay the polling message to the next terminal if it has nothing to transmit. This polling method requires more logic in the terminal because it requires the terminal to change its action when its normal addressee (the next terminal) is inoperative.

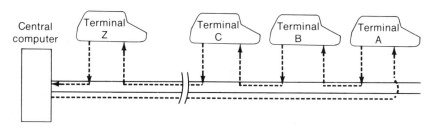

Figure 4-3:   Hub-Go-Ahead Polling

Figure 4-4 shows an example of a roll call polling scheme implemented by means of a remote controller. Roll call polling allows assigning priority to a specific terminal address, say (2). This might be more advantageous than hub-go-ahead polling because roll call polling can give higher priority to the line printer (2), which may have a high volume of data. Hub-go-ahead polling will give equal priority to all devices, (1) through (4).

Because response time requirements and traffic volumes can change throughout a day, it may be advantageous to change the polling list within the CPU, front-end communications processor, or concentrator. There may be peak volumes of message transmissions from various parts of the country at some times during the business day, and, in this case, the network designer may develop two or more different polling lists that offer different priorities to terminal locations in various geographic regions. One polling list may not meet the throughput requirements of a network during the business day. Also, when a terminal is out of service the polling list should be modified so that the terminal cannot be polled. Polling an inoperative terminal wastes time and offers an opportunity for illegal entry into the data communication network by unauthorized personnel.

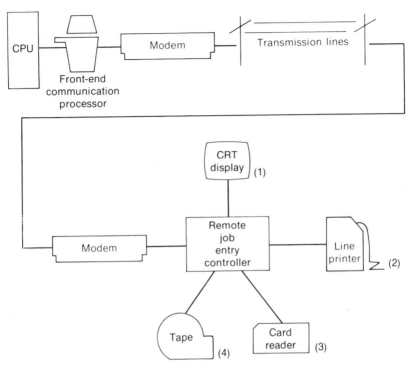

Figure 4-4: Polling a Remote Job Entry Station by the
Roll Call Method

Line Protocol

Line protocol is the term given to the exchange of predetermined
signals for purposes of control when a connection is to be established between a terminal and a central computer. This exchange
of signals between data communications equipment during the
message setup period is sometimes called "handshaking." It includes the verification that the called and calling devices are the
correct ones, establishes one device as the master (controller of
the call) and one as the slave, it determines the direction of the
initial transmission, and determines the mode in which the terminal is currently operating such as idle mode or a control mode.
When in a control mode, the terminal is ready to accept or
transmit a message.

An example of a line protocol is shown in Figure 4-5. In this
figure, a polling sequence is depicted and the line protocol procedures are shown step by step. This example represents a simple,
basic form of terminal computer interconnection. Notice, from

Steps	Central Computer Site	Remote Terminal Site
1.	Poll Terminal A	————————▶
2.		◀———————— Negative Response Code
3.	Poll Terminal B	————————▶
4.		◀———————— Positive response code (changes terminal mode from idle to text mode)
		◀———————— Text (first block of text)
		◀———————— End of block code
		◀———————— Parity check
5.	Parity received correctly ————————▶	
6.		◀———————— Text (second block of text)
		◀———————— End of block code
		◀———————— Parity check
7.	Parity received correctly ————————▶	
8.		◀———————— End of message
9.	Poll Terminal C	————————▶
10.		◀———————— Negative response code

Figure 4-5: Line Polling Protocol

the point that the computer polls Terminal B, there are five line turnarounds until the end of message transmission by the remote terminal site. Each of these line turnarounds consumes time (modems/terminals typically require 30 to 80 milliseconds and echo suppressors on voice grade lines require 150 to 300 milliseconds). Therefore, a line protocol that reduces the number of line turnarounds is very desirable. Half-duplex transmission on a four-wire circuit is an effective way to reduce line turnaround time because the messages in each direction are transmitted on a separate pair of wires, thereby eliminating echo suppressor turnaround time.

Another technique that eliminates much of the excess turnaround is to allow the next message to carry the acknowledgement of the previous message. If a message is not ready, a specially prepared acknowledgment message is sent. For example, a message may be sent from the central computer to a remote terminal, or vice versa, utilizing the message protocol format shown in Figure 4-6. Each transmission has a "start-of message" character (SOH) which is followed by the address character (ADD) for the terminal to which the message is being sent. This system utilizes the "acknowledge-receipt-of-message" character

```
            →
B   E                    S   T   T   A   A   S
P   T        (TEXT)      T   R   R   C   D   O
C   X                    X   N   N   K   D   H
```

SOH — Start of message
ADD — Device (terminal) address
ACK — Acknowledge receipt of message
TRN — Transmission number (2 digits)
STX — Start of text
ETX — End of text
BPC — Block parity check

Figure 4-6: Message Protocol Format

(ACK) and the two-digit transmission number characters (TRN) to acknowledge prior messages by keeping track of the message numbers between the terminal and the computer. The TRN is increased by one for each successful transmission of a message. The ACK character is the low order digit of the transmission number of the most recent successfully received message. Thus, messages may be acknowledged by another message (if one is ready for transmission).

Figure 4-7, Message Protocol Scheme, depicts how this transmission scheme operates. Each message uses the format of Figure 4-6. In Figure 4-7, the message transmission scheme goes from 1 to 2 to 3 to 4. Also, the reader must remember that the data is being transmitted in blocks that are made up of a start of message character (SOH), a device address character (ADD), and acknowledgement character (ACK), a transmission number (TRN), a start of text character (STX), the text or message itself, the end of text character (ETX), and finally the block parity check character (BPC).

As you review Figure 4-7, you will notice that in the last transmission (Item 4 in lower right corner) the acknowledgement shows the unsuccessful receipt of Message 02 because the ACK character was not incremented from a 1 to a 2. In this case, when Station B transmits its next message back to Station A, Station A will notice that the ACK number is low and will retransmit Message 02. If Station B does not have another message to send, then it will send a message block that does not contain any text but does increment the TRN code and the ACK code.

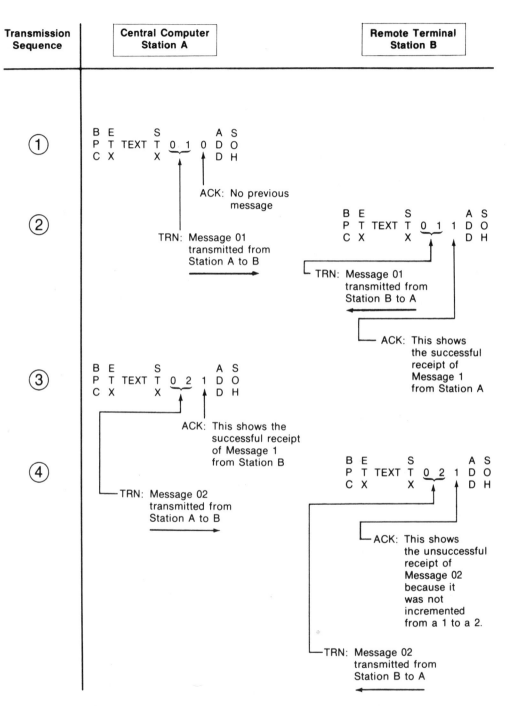

Figure 4-7:   Message Protocol Scheme

Switched Versus Non-Switched Networks

Whenever data are to be sent within a network having many terminal locations, some arrangement must be made to enable different terminals to communicate with each other. Early communication systems utilized permanent connections between each pair of terminals. When the number of terminals exceeds two or three, this gets increasingly expensive. Figure 4-8 depicts this situation; it shows six terminals, with every station individually connected to every other station. Fifteen lines are needed in this case. The number of lines needed for a network of this type may be calculated as follows:

$$\text{Number of lines} = \frac{N^2 - N}{2} \quad \text{where}$$

N = the number of terminals or stations.

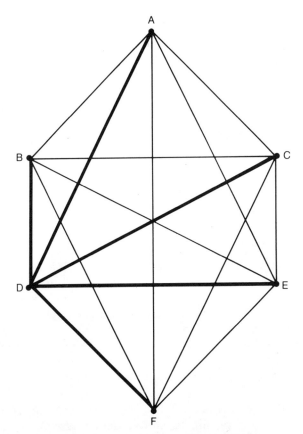

Figure 4-8:   Direct Wire Connection for Six Terminals

A more economical and flexible type of interconnection arrangement is to allow for the temporary connection between any two stations that wish to communicate with each other. This process is called switching. This methodology eliminates the need for direct wire connection between all station pairs in a network. If Station D had the ability to temporarily connect any two stations, then the 15-line configuration of Figure 4-8 could be reduced to a 5-line configuration, consisting of one line from each station going into the central switching station (D).

The basic functions of a switching system are:

- Interconnecting the various stations to provide a communication path between the stations.
- Controlling the establishment and release of the various connections.
- Checking network equipment to determine whether it is busy or inoperative.

There are two types of switched networks, line-switched networks and store-and-forward switched networks. Line-switched networks work in real time. In line switching, the central site establishes a connection between two stations and a message goes directly from one terminal to another. Store-and-forward switched networks have the capability, at the switch point, to receive, store, and subsequently transmit messages sent on the network. The most common and important line-switched network is the public dial telephone network. An understanding of its basic operation is important to the data communication system designer and user.

Store-and-Forward Networks

In the store-and-forward type of communication switching system, the central switching site stores the incoming message from the sending terminal by copying it onto a storage medium such as paper tape, magnetic tape, or disk and, at a later time, retransmits that message to the destination terminal.

Typical of the older type of store-and-forward switching system is the manual "torn tape" switch. In this system, Station A transmits a message to the switching center. The message is received on punched paper tape and torn off the paper tape perforator that is connected to Station A. At this point, the message may or may not be stored depending on whether the station to which it is addressed is busy or free. If the station is free, the punched paper tape is placed in the paper tape reader that is connected to the station to which the message is addressed.

With the advent of modern high-speed computers, data communication networks are able to combine the store-and-forward concept with the circuit or line-switching concept. This combination offers data communication network users the highest level of throughput. A modern communication switching network will offer circuit or line switching, so the originator of a message can be immediately connected to the station to which the message is addressed; but if that station is busy, the computer center will accept the message, store it on a magnetic device (generally disk), and automatically transmit the message to the proper addressee as soon as the line becomes free.

Each category of communication switching systems has its advantages and disadvantages. The tape switching systems are being phased out because of the higher efficiency of circuit or line switching, with the accompanying advantages of computer controlled store-and-forward capabilities. Circuit or line switching is characteristic of the normal telephone transmission system and is usually required for on-line real-time data communication systems. It is apparent that in large information systems, a combination of circuit or line switching and store-and-forward techniques will have to be used to provide the necessary data communication throughput.

The Dial Telephone System

The telephone system in the United States is dominated by American Telephone and Telegraph Company (the Bell System) which has over 80% of all telephones installed. The second largest telephone company is General Telephone and Electronics with approximately 8% of all telephones installed. About 1600 other independent telephone companies share the remaining 10% of the telephone business. Operation of the system is made possible because all these companies are connected to the toll trunk network that allows interconnection between the various companies. It is this telephone network system that carries most of the data communication data traffic between the various business organizations that lease telephone lines or circuits. The telephone network is a line switching system.

The basic components of the dial telephone system are depicted in Figure 4-9. The telephone instrument, often called a station, is depicted in the figure by the telephone symbols lettered

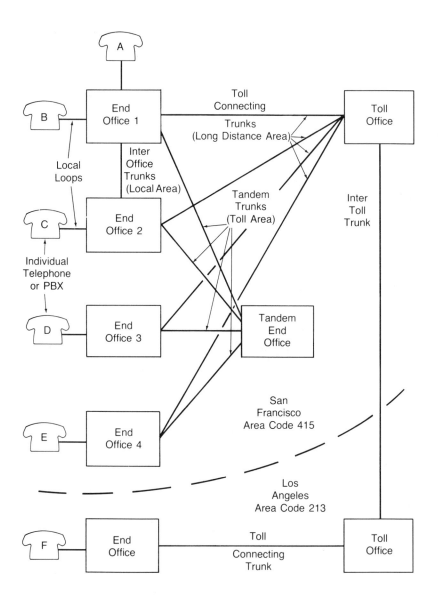

Figure 4-9:   The Dial Telephone System

A through F. These stations can be an individual's telephone, a switchboard (a private branch exchange, or **PBX**), or an organization's input/output terminal utilized for data communications traffic. The end office (also called central office) is the switching center which, through the use of electronic and/or electromechanical equipment, automatically switches calls to the desired party. The tandem end office is a facility that switches calls from one end office to another end office within the same geographical area. A toll office is a facility that controls the incoming or outgoing calls for a particular region, such as an area code in the telephone dialing scheme.

A local call is a call placed by one party to another party, both of whom are connected to the same end office, such as Stations A and B. Sometimes, in large metropolitan areas with high population densities, a few end offices may be directly connected, such as End Offices 1 and 2 in the figure. In this case, when Station B calls Station C it is also considered a local call.

A toll (message unit) call is a call placed by one subscriber to another subscriber, where the called subscriber is in the same area code or geographical area as the caller but is outside the caller's free calling area. In this case, the call must go through the tandem end office. For example, a call from Subscriber B to Subscriber E would be a toll call. The tandem end office must perform the interconnection between End office 1 and End office 4.

A long distance call is a call placed by a subscriber in one area code to another subscriber in a different area code. Long distance calls go through the toll office of each area code. For example, when Subscriber B calls Subscriber F, it is considered a long distance call.

The local end office, the tandem end office, and the toll office are known as switching centers. These switching centers detect that service is required when a subscriber picks up the telephone. These switching centers also set up the interconnection path between the called and calling parties, make the telephone of the called party ring, detect busy conditions, discontinue the telephone ringing once the called party answers, disconnect the circuit upon completion of the call, and charge the caller for the telephone call if it is either a toll call or a long distance call. Because of the logic used in the switching centers, different pathways may be used at different times to connect the same two subscribers. When a calling subscriber receives a busy signal of 60 impulses per minute, it means that the called subscriber's telephone is busy (i.e., this signal comes from the called party's end office) but, if the busy signal is 120 impulses per minute, that means that the intertoll trunk is busy (i.e., the area code is busy).

There are various supervisory signals on dial telephone lines with which the data communication functions must not interfere. This signaling is present only when the data communication system is using the dial-up telephone network. When private leased lines are utilized, this network signaling is not present. These supervisory signals include dialing signals, which give the address of the end office, tandem office, toll office, and the specific telephone subscriber that is being called, and status signaling, such as line busy and control signals that perform various operations on equipment, such as coin return on a pay telephone.

A Private Branch Exchange (PBX) is a switchboard within a subscriber's organization where all the telephone lines of the organization terminate. Usually, several lines go from this switchboard to an end office. The line between an end office and an individual subscriber or an organizational subscriber is called a local loop. This is the telephone line between business premises and the local telephone company end office that services the business premises. A business organization might have a dozen incoming telephone lines (local loops) going to a PBX, and the PBX might have 200 telephone extensions connected to it. When a call comes into the PBX switchboard, the operator determines the extension of the called party and then manually plugs the call into the correct extension and rings that telephone. When that person answers, a busy light comes on signifying that the line is busy. The busy light goes off as soon as either party hangs up; the operator then removes the plug and that extension is available for use again. When a PABX (an automatic PBX) is used, operator intervention is not required, in contrast to a regular PBX.

Centrex is the Bell System trade name for an automatic system that can route incoming or outgoing calls directly to the person being called without the intervention of a PBX operator. Each station has its own telephone number, the last four digits of which are the internal organizational extension number. With Centrex, incoming calls go directly to the station being called and are not intercepted by a PBX operator. The stations can dial inside or outside calls, directly, thus bypassing the PBX operator. Centrex does not totally eliminate the PBX operator because there still must be a central organizational telephone number for general information purposes.

Questions—Chapter 4

This chapter has explained the various ways to interconnect network users, and disciplines that must be imposed to provide for the orderly use of system resources. This is a key aspect of data communications system design. When dozens, or even hundreds, of users are contending with one another to have their communication needs satisfied, the primary determinant of service level is the quality of the network design. As in earlier chapters, there are four types of questions: True or False, Fill-In, Multiple Choice, and Short Answer. The first four short answer questions require analysis and generalization; the remainder require fact feedback.

True or False

1.   Point-to-point lines have two terminals.

2.   Switched networks are built of point-to-point multidrop lines.

3.   The most common form of polling is hub-go-ahead polling.

4.   Hub-go-ahead polling reduces the number of line turnarounds necessary on a multidrop line.

5.   If there are peak volumes of message transmissions from certain terminals during only a part of the day, it may be advantageous to develop more than one polling list.

6.   When in control mode, a terminal is unable to accept or transmit a message because it is under the control of another device.

7.   ACK and TRN characters are used to reduce the number of line turnarounds and facilitate the detection of a lost message.

8.   If a separate line was used to connect each pair of terminals in a communication system with 10 terminals, 45 lines would be required.

9.   The telephone network is a line switching network.

10.   A "torn tape" switch refers to a malfunction condition that infrequently occurs using magnetic tape in a modern store-and-forward network.

Fill-in

1.   Information routings change with time in a ——— network.

2.   In order for a terminal to receive a message on a multidrop line, each terminal must have its own ——— .

3. ——— addresses involve the ability to send a message down a multidrop line with an address than can be recognized by all terminals sharing that line.

4. ——— ——— are a form of multidrop line which is very efficient when several terminals are located in close proximity.

5. A ——— system is a data communication network in which terminals seize network facilities on a first-come, first-served basis.

6. ——— ——— is the term given to the exchange of predetermined signals for purposes of control when a connection is to be established between a terminal and a central computer.

7. The process of temporarily connecting any two stations is called ——— .

8. There are ——— basic types of switched networks.

9. In ——— ——— the central site only establishes a connection between the two stations and the message goes directly from one terminal to the other.

10. PBX stands for ——— ——— ——— .

Multiple Choice

1. Handshaking includes which of the following functions?
    a) Determining the direction of the initial transmission
    b) Determining whether the terminal is in an idle mode
    c) Determining which is the master and which is the slave
    d) All the above
    e) None of the above

2. A line turnaround for echo suppressers on voice grade lines requires approximately
    a) 30 to 80 microseconds
    b) 30 to 80 milliseconds
    c) 150–300 milliseconds
    d) 960–5000 bits per second
    e) None of the above

3. Which of the following is the type of office that a station would be connected to first:
    a) End office
    b) Intermediary office

c) Tandem office
d) Toll office
e) None of the above

4. The two biggest telephone companies together have approx-
    imately what percentage of all telephones installed in the
    United States?
    a) 19%
    b) 50%
    c) 88%
    d) 95%
    e) 99.9%

5. Which of the following is the first character transmitted in
    the message protocol scheme illustrated in the text?
    a) ACK
    b) ADD
    c) SOH
    d) STX
    e) None of the above

6. Which of the following is the last character transmitted in
    the message protocol scheme illustrated in the text?
    a) ACK
    b) ADD
    c) SOH
    d) STX
    e) None of the above

7. Which of the following characters would immediately fol-
    low the text portion of the message sent using the message
    protocol scheme illustrated in the text?
    a) ACK
    b) ADD
    c) SOH
    d) STX
    e) None of the above

Short Answer

1. Make a list that can be used to compare the characteristics
    of various forms of non-switched network building blocks.
    Select a set of characteristics that will serve to point up
    differences, advantages, and disadvantages of point-to-point
    and multidrop lines.

2.   Using Figure 4-5 as a model, invent a line protocol for sending a message to the terminal.

3.   Find a PBX and watch its operation for a few minutes. Obtain from the telephone operator and list the following data from it: number of incoming lines, number of stations, number of incoming calls per hour (average and peak), number of outgoing calls (average and peak), average outside call duration, and type of PBX (manual or automatic). Calculate the percentage utilization of the incoming lines.

4.   What does the following describe? The central computer sends a message saying, "Terminal A, do you have anything to transmit? If so, send it." If Terminal A has nothing to send, it replies negatively and the central computer goes on to Terminal B.

5.   What is a specific type of polling that is used on multidrop lines?

6.   What is a line protocol?

7.   Can line switching and store-and-forward switching be combined in the same system? If so, describe how it would work.

8.   What is an end office?

5

Error Detection and Correction

So far in this book, we have dealt largely with methods
and equipment that we have assumed to be fault
free. This is not the case in real-world situations. In most
data communication systems, the control of errors is vitally
important. There are many causes of errors as well as a
variety of approaches that a systems designer
should consider in dealing with error detection and
correction procedures. This chapter equips the
systems designer to handle data communication errors through
an understanding of line noise, approaches to error
control, error correction, and data blocking.

Data Communication Errors

Errors are a fact of life in data communications. Depending on
the type of line, they may occur every few minutes or every few
seconds or even more frequently. They occur because of noise on
the lines (types of line noise are discussed in the next section). No
data communication system can prevent all these errors from
occurring, but most of them can be detected and many corrected
by proper design. Common carriers that lease data transmission
lines to users provide statistical measures specifying typical error
rates and the pattern of errors that can be expected on the
different types of lines they lease.

Normally, errors appear in bursts. In a burst error, more than one data bit is changed by the error-causing condition. This is another way of saying that 1-bit errors are not uniformly distributed in time. However, common carriers usually list their error rates as the number of bits in error divided by the number of bits transmitted, without reference to their non-uniform distribution. For example, the error rate might be given as 1 in 500,000 when transmitting on a public voice grade telephone line at 1,200 bps.

The fact that errors tend to be clustered in bursts rather than evenly dispersed has both positive and negative aspects. If the errors were not clustered (but instead were evenly distributed throughout the day) with an error rate of 1 bit in 500,000, it would be rare for two erroneous bits to occur in the same character, and consequently some simple character checking scheme would be effective. But this is not the case, because bursts of errors are the rule rather than the exception. They sometimes go on for time periods that may obliterate 50 to 100 or more bits. The positive aspect is that, between bursts, there may be rather long periods of error-free transmission. Therefore no errors at all may occur during data transmission in a large proportion of messages. For exa#ple, when errors are #ore or less evenly distrib#ted, it is not di#ficult to gras# the me#ning even when the error #ate is high, as it is in this #entence (1 charac#er in 20). On the other hand, if errors are concentrated in bursts, it becomes more difficult to recover the meaning and much more reliance must be placed on knowledge of message #######* or on special logical/numerical error detection and correction methods.

It is possible to develop data transmission methodologies that give very high error detection and correction performance. The only way to do the detection and correction is to send along extra data. The more extra data that are sent, the more error protection that can be achieved. But as this protection is increased, the throughput of useful data is reduced. Therefore, the efficiency of data throughput varies inversely as the desired amount of error detection and correction is increased. Errors will even have an effect on the length of the block of data to be transmitted when using synchronous transmission. The shorter the message blocks used, the less likelihood there is of needing retransmission for any one block. But the shorter the message block, the less efficient is the transmission methodology as far as throughput is concerned. If the message blocks are long, a higher proportion may have an error and have to be resent.

---

*In case you could not guess, the word is "context."

In transmissions over the public switched network, a considerable variation in the error rate is found from one time of the day to another. The error rate is usually higher during the periods of high traffic (the normal business day). In some cases, the only alternative open to the user of these facilities is to transmit the data at a slower speed because higher transmission speeds are more error prone. Dial-up lines are more prone to errors because they have less stable transmission parameters than private leased lines, and, because different calls use different circuits, they usually experience different transmission conditions. Thus, a bad line is not necessarily a serious problem in dial-up transmission; a new call may result in getting a better line. Line conditioning is not available on dial-up lines. Line conditioning, a service that is available only on private leased lines, consists of special electrical balancing of the circuit to ensure the most error-free transmission.

Line Noise and Distortion

Line noise and distortion can cause data communication errors. In this context we define noise as undesirable electrical signals. It is introduced by equipment or natural disturbances and it degrades the performance of a communication line. If noise occurs, the errors are manifested as extra or missing bits, or bits whose states have been "flipped," with the result that the message content is degraded. Line noise and distortion can be classified into roughly 10 categories, including white noise, impulse noise, cross talk, echoes, intermodulation noise, amplitude changes, line outages, attenuation, delay distortion and jitter.

White or Gaussian noise is the familiar background hiss or static on radio and telephones. It is noise caused by the thermal agitation of electrons and because of this, it is inescapable. Even if the equipment utilized were perfect and the wires were perfectly insulated from any and all external interference, there would still be some white noise. White noise is usually not a problem unless its level becomes so high that it obliterates the data transmission. Sometimes noise from other sources such as power line induction, cross modulation from adjacent lines, and a conglomeration of random signals resembles white noise and is labeled as such even though it is not caused by thermal electrons.

Impulse noise (sometimes called spikes) is the primary source of errors in data communications. An impulse of noise can last as long as 1/100th of a second. An impulse of this duration would be heard as a click or a crackling noise during voice communications. This click would not affect voice communications but it might obliterate a group of data bits causing a burst error on a data communication line. At 150 bps, 1 or 2 bits would be changed by a spike of 1/100th of a second, while at 4,800 bps, 48 bits would be changed. Some of the sources of impulse noise are voltage changes in adjacent lines or circuitry surrounding the data communication line, telephone switching equipment at the telephone exchange branch offices, arcing of the relays at older telephone exchange offices, tones used by network signaling, maintenance equipment during line testing, lightning flashes during thunder storms, and intermittent electrical connections in the data communication equipment.

Cross talk occurs when one line picks up some of the signal traveling down another line. Cross talk occurs between line pairs that are carrying separate signals, in multiplexed lines carrying many discrete signals, in microwave links where one antenna picks up a minute reflected portion of the signal from another antenna on the same tower, and in any hard-wire telephone circuits that run parallel to each other, are too close to each other, and are not electrically balanced. You are experiencing cross talk during voice communication on the public switched network when you hear other conversations in the background. Cross talk between lines will increase with increased communications distance, increased proximity of the two wires, increased signal strength, and higher frequency signals. Cross talk, like white noise, has such a low signal strength that it is normally not bothersome on data communication networks.

Echoes and echo suppression can be a cause of errors (echo suppressors were discussed in Chapter 2). An echo suppressor causes a change in the electrical balance of a line and this change causes a signal to be reflected so it travels back down the line at reduced signal strength. Whenever, as in data transmission, the echo suppressors are disabled, this echo returns to the transmitting equipment. If the signal strength of the echo is high enough to be detected by the communication equipment it will cause errors. Echoes, like cross talk and white noise, have such a low signal strength that they are normally not bothersome.

Intermodulation noise is a special type of cross talk. The signals from two independent lines intermodulate and form a product that falls into a frequency band differing from both inputs. This resultant frequency may fall into a frequency band that is reserved for another signal. This type of noise is similar to harmonics in music. On a multiplexed line, many different signals are amplified together and slight variations in the adjustment of the equipment can cause intermodulation noise. A maladjusted modem may transmit a strong frequency tone when not transmitting data, thus yielding this type of noise.

Amplitude noise involves a sudden change in the level of power. The effect of this noise depends on the type of modulation being used by the modem. For example, amplitude noise does not affect frequency modulation techniques. This is because the transmitting and receiving equipment interprets frequency information and disregards the amplitude information. Some of the causes of amplitude noise may be faulty amplifiers, dirty contacts with variable resistances, sudden added loads by new circuits being switched on during the day, maintenance work in progress, and switching to different transmission lines.

Line outages are a catastrophic cause of errors and incomplete transmission. Occasionally a communication line fails for a brief period of time. This type of failure may be caused by faulty telephone branch office exchange equipment, storms, loss of the carrier signal, and any other failure that causes an open line or short circuit.

Attenuation is the loss of power that the signal suffers as it travels from the transmitting device to the receiving device. It results from the power that is absorbed by the transmission medium or lost before it reaches the receiver. As the transmission medium absorbs this power, the signal gets weaker, and the receiving equipment has less and less chance of correctly interpreting the data. To avoid this, telephone lines have repeaters (also called amplifiers) spaced throughout their length. The distance between them depends upon the amount of power lost per unit length of the transmission line. This power loss is a function of the transmission method and medium. Also, attenuation increases as frequency increases or as the diameter of the wire decreases.

Attenuation distortion is where high frequencies lose power more rapidly than low frequencies during transmission. This can cause the received signal to be distorted by unequal loss of its component frequencies.

Delay distortion can cause errors in data transmission. Delay distortion ocurs when a signal is delayed more at some frequencies than at others. If the method of data transmission involves data transmitted at two different frequencies, then the bits being transmitted at one frequency may travel slightly faster than the bits transmitted at a different frequency. A piece of equipment, called an equalizer, compensates for both attenuation distortion and delay distortion.

Jitter may affect the accuracy of the data being transmitted. The generation of a pure carrier signal is impossible. Minute variations in amplitude, phase, and frequency always occur. Signal impairment may be caused by continuously and rapidly changing gain and/or phase changes. This may be random or periodic and is defined as jitter.

Approaches to Error Control

Error control implies (1) techniques of design and manufacture of data communication transmission links and equipment to reduce the occurrence of errors (an area that is outside the scope of this book), and (2) methodologies to detect and correct the errors that are introduced during transmission of the data. In the sense of the second meaning of error control, the methodologies fall into four categories:

- Ignoring errors
- Loop or echo checking
- Error detection with retransmission
- Error detection with automatic correction

Each of these approaches will be discussed in this section.

Ignore the Errors

The first of the approaches is to ignore the errors. This is not as foolish as it may initially seem. It can be done in data communication systems that handle noncritical messages made up primarily of English language text or other redundant expression methods and that do not rely upon accurate numerical values.

When a telegram is sent, no error detection or correction techniques are employed. The methodology of ignoring the errors in a telegram works because people can usually read bad text when only a few characters are in error. (Remember the example earlier in this chapter.) This methodology reduces the costs of data transmission and increases the throughput because no error detection and correction schemes are involved. This is generally not a viable alternative for organizations such as businesses and governments because accuracy is usually a prerequisite.

Loop or Echo Checking

Loop or echo checking does not use a special code. Instead, each character, or other small unit of the message, as it is received is transmitted back to the transmitter, which checks to determine whether the character is the same as the one just sent. If it is not correct, then the character is transmitted a second time. This method of error detection is wasteful of transmission capacity because each message (in pieces) is transmitted at least twice and there is no guarantee that some messages might not be transmitted three or four times. Also, some of this retransmission of characters for a second or third time might not be necessary since the error could have occurred on the return trip of the character. This would require the transmitter to retransmit the character even though it was, in reality, received correctly the first time. Loop or echo checking is usually utilized on hard-wire, short lines, with low-speed terminals. This type of error checking does give a certain degree of protection but it is not as efficient as other methods.

Error Detection with Retransmission

Error detection and retransmission schemes are built into data transmitting and receiving devices, front-end computers, modems, and software. These schemes include detection of an error and immediate retransmission, detection of an error and retransmission at a later time, or detection of an error and retransmission for up to, say, three tries and then retransmission at a later time, or the like. Error detection and retransmission is the simplest, and if properly handled, the most effective and least expensive method to reduce errors in data transmission. It requires the simplest logic, relatively little storage, is best understood by terminal operators, and is most frequently used. Retransmission of the message in error is straightforward. It is usually called for by the failure of the transmitter to receive a positive acknowledgement within a preset time. Various methods are used to determine that the message that has just been received has, in fact, an error imbedded in it. Some of the common error detection methods are parity checking, constant ratio codes, and polynomial checking.

Parity checking: Upon examining a common form of the USASCII coding structure, it soon becomes apparent that 1 of the 8 bits encoding each character is redundant—i.e., its value is solely determined by the values of the other 7 and is therefore unnecessary. Since this 8th bit cannot transmit any new information, its purpose is the confirmation of old information. The logic of its use is shown in Figure 5-1. The most common rule for fixing the value of the redundant bit uses the "parity" (evenness or oddness) of the number of 1's in the code. Thus, for an even parity code system using USASCII:

- Letter "V" is encoded 0110101 and a zero is added in the parity (8th) position since the number of 1's is 4, already an even number, yielding V=01101010

- Letter "W" is encoded 0001101 and a 1 is added in the parity position to make the number of 1's even, yielding W=00011011.

A little thought will convince you that any single error (a switch of a 1 to a 0 or vice versa) will be detected by a parity check but that nothing can be deduced about which bit was in error. If the states of two bits are switched, the parity check may not sense any error. Of course, it may be possible to sense such an

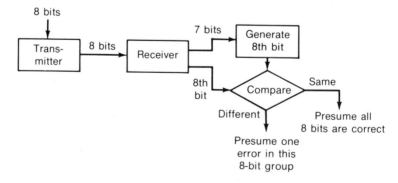

Figure 5-1:   Parity Checking Logic

error because the resulting code, although correct as far as parity is concerned, is a code that is "forbidden," e.g., undefined or inappropriate in its context. Such detection, of course, requires more circuitry or software.

What is the cost of the parity bit? It consists of two parts: the cost of the generating and comparison circuits, and the cost of channel capacity. Recent technology has reduced the costs of the circuitry to low levels. The cost of channel capacity, however, is increased by the need to send 8 bits to get 7 bits of "real" information sent. In other words, the code we have been examining is $7/8 = 0.875$ efficient. If we sent this code on a 2,400 bps channel, the "real" information transfer rate would be $0.875 \times 2,400 = 2,100$ bps.

This is only one way that information transfer rates are diluted by the realities of data communications. Another source of dilution is the transmission technique. The USASCII code is commonly used in Teletype operations. In that case, it is sent asynchronously i.e., without a time base that is common to sender and receiver. The individual bits making up one character are fairly well synchronized by the mechanics of the Teletype transmitter, but the timing between characters is completely arbitrary (especially if a "hunt-and-peck" typist is at the keyboard). Thus, as was discussed in Chapter 2, some provision must exist to help the receiver identify a character as it arrives and tell when a character is complete. To do this for Teletypes, three more 1-bits are added to the 8-bit character, one in front and two following the character. Now the efficiency is $7/11 = 0.636$, and asynchronous transmission over a 110-bps line will yield only 70 bps or 10 characters of "real" information per second. The alternative to asynchronous transmission (synchronous transmission) also suffers from a related dilution problem which will be discussed later.

The example above deals only with the simplest form of parity checks. Figure 5-2 contrasts two different parity checking techniques. Utilizing the example of punched paper tape, as shown in Figure 5-2, a vertical parity check is a parity on a single character. A combination of vertical and horizontal parity checking checks each character's parity and each horizontal row of data bits for parity. In other words, there is one parity bit for each character and a parity character for the total of all characters in a message block.

Still another parity checking technique is the cyclical parity check (sometimes called interlaced parity). This method requires two parity bits per character. Assuming a 6-data bit code structure, the first parity bit would provide parity for the first, third,

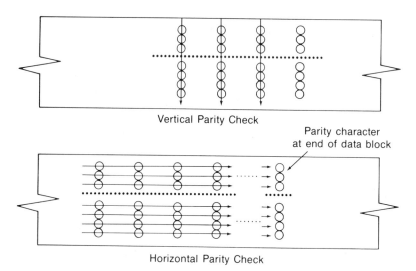

Vertical Parity Check

Parity character
at end of data block

Horizontal Parity Check

Figure 5-2: Parity Checking Techniques

and fifth bits, and the second parity bit would provide parity for the second, fourth, and sixth bits. Figure 5-3 shows an even parity cyclical parity check on a 6-bit code.

Constant ratio codes Constant ratio codes are special data communication codes that have a constant ratio of the number of 1 bits to the number of 0 bits. The most common constant ratio code is IBM's 4-of-8 code that was discussed in Chapter 2. Constant ratio codes detect an error whenever the number of 1 bits and 0 bits are not in their proper ratio. For example, in the 4-of-8 code there are always supposed to be four 1 bits and four 0 bits in the received bit configuration of the character. Whenever this ratio is out of balance the receiving equipment knows that an error has occurred. Constant ratio codes are not widely utilized because they are inefficient. As an example of their inefficiency, consider the 4-of-8 constant ratio code, which has 70 valid character combinations while a 7-bit USASCII code has 128 valid character combinations ($2^7 = 128$).

Polynomial checking Polynomial checks on blocks of data are often performed for synchronous data transmission. In this type of message checking, all the bits of the message are checked by applying a mathematical algorithm. For example, all the 1 bits in a message are counted and then divided by a prime number (such as 17) and the remainder of that division is transmitted to the

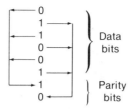

Figure 5-3: Cyclical Parity Check for a 6-Bit Code

receiving equipment. The receiving equipment performs the same mathematical computations and matches the remainder that it calculated against the remainder that was transmitted with the message. If the two are equal, the entire message block is assumed to have been received correctly. In actual practice, much more complex algorithms are utilized.

Error Detection with Automatic Correction

This approach, sometimes called forward error correction, involves codes that contain sufficient redundancy to permit errors to be detected and corrected at the receiving equipment without retransmission of the original message. The redundancy, or extra bits required, varies with different schemes. It ranges from a small percentage of extra bits to 100% redundancy, with the number of error-detecting bits roughly equaling the number of data bits. One of the characteristics of many error-correcting codes is that there must be a minimum number of error-free bits between bursts of errors. For example, one such code, called a Hagelbarger Code, will correct up to six consecutive bit errors provided that the 6-bit error group is followed by at least 19 valid bits before further error bits are encountered. Bell Telephone engineers have developed an error correcting code that uses 12 check bits for each 48 data bits, or 25% redundancy. Still another code is the Bose-Chaudhuri Code that, in one of its forms, is capable of correcting double errors and can detect up to four errors.

To show how such a code works, consider this example of a forward error-correcting code, called a Hamming Code, after its inventor, R. W. Hamming.* This code associates even parity bits

*William P. Davenport, *Modern Data Communication-Concepts, Language, and Media* (New York: Hayden Book Company, Inc., 1971), p. 96.

with unique combinations of data bits. Using a 4-data-bit code as an example, a character might be represented by the data bit configuration 1010. Three parity bits P_1, P_2, and P_4 are added, resulting in a 7-bit code, as shown in the upper half of Figure 5-4. Notice that the data bits (D_3, D_5, D_6, D_7) are 1010 and the parity bits (P_1, P_2, P_4) are 101.

As depicted in the upper half of Figure 5-4, parity bit P_1 applies to data bits D_3, D_5, and D_7. Parity bit P_2 applies to data bits D_3, D_6, and D_7. Parity bit P_4 applies to data bits D_5, D_6, and D_7. For the example, in which D_3, D_5, D_6, D_7 = 1010, P_1 must equal 1 since there is but one 1 among D_3, D_5, and D_7 and parity must be even. Similarly P_2 must be 0 since D_3 and D_6 are 1's. P_4 is 1 since D_6 is the only 1 among D_5, D_6, D_7.

Now, assume that during the transmission, data bit D_7 is changed from a 0 to a 1 by line noise. Because this data bit is being checked by P_1, P_2, and P_4, all three parity bits will now show odd parity instead of the correct even parity. (D_7 is the only data bit that is monitored by all three parity bits, therefore whenever D_7 is in error all three parity bits will show an incorrect parity.) In this way, the receiving equipment can determine which bit was in error and reverse its state, thus correcting the error without retransmission.

The bottom half of Figure 5-4 is a table that determines the location of the bit in error. A 1 in the table means that the corresponding parity bit indicates a parity error. Conversely, a zero means the parity check is correct. These 0's and 1's form a binary number that indicates the numerical location of the erroneous bit. In the example above, P_1, P_2, and P_4 checks all failed, yielding 111, or a decimal 7, the subscript of the erroneous bit.

Error detection and correction methodologies come in many varieties. The data communication network designer must give careful consideration to all facets of the system being designed and make appropriate use of the various methodologies to control and correct errors.

Encryption

Encryption is the process of transforming a message that is in understandable form, ("Plain Text") into an equivalent message that is in a form not immediately understandable ("Cipher Text"). The reverse process is called decryption. Encryption differs from encoding in that it works on the basic message to increase privacy, rather than to facilitate transmission.

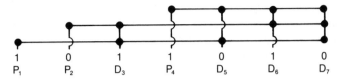

Checking Relations Between Parity Bits (P) and Data Bits (D)

0 = Corresponding
   parity check is correct
1 = Corresponding parity
   check fails
			Determines in
			which bit the
			error
			occurred

P_4	P_2	P_1	
0	0	0	→ no error
0	0	1	→ P_1
0	1	0	→ P_2
0	1	1	→ D_3
1	0	0	→ P_4
1	0	1	→ D_5
1	1	0	→ D_6
1	1	1	→ D_7

Interpreting Parity Bit Patterns

Figure 5-4:   Hamming Code for Forward Error Correction

Encryption can operate on an analog signal, and is then known as "Scrambling." The most important process for the data communication systems designer, however, is the logical encryption of digital data. This may be accomplished by special-purpose, hard-wired devices or by programmable devices, or by a combination of the two.

There are many transformation algorithms for encryption, some of which date back hundreds of years and all of which were inspired by the need for privacy in political and military communications.* In the environment of today's applications for data communications, encryption is becoming more important. It is not a simple subject since it requires higher mathematics for an understanding of the relative privacy of transformations and their techniques of use. If the system designer is confronted with a real

---

*A fascinating history and tutorial on this subject is David Kahn, *The Codebreakers* (New York: Macmillan Co., 1967).

need for substantial privacy in a data communication system, detailed research into techniques and alternatives, or the advice of a specialist are the best paths to follow.

Human Factors in Error Control

Many of the errors in data communication systems are caused by humans, rather than being introduced by equipment or during transmission over the communication links. Operator performance is a much overlooked area of systems engineering. The best-engineered equipment can be only as good as the performance of the operator.

In error prevention, which is an excellent way to achieve error control, the data communication systems designer should consider the following items:

- Adequate operator training should be provided. Operator manuals, short-form procedures using mnemonics or numbers for experienced operators, and terminal prompting will assist in keeping the error rate low among experienced operators.

- An uncomplicated dialog between the operator and the applications system should be developed. A simple keyboard (special numeric keyboards may be installed if numbers are used frequently), and the inclusion of features for control of errors such as an erase key, backspace key, etc., will ease the burden of work and result in fewer errors.

- Instructions should be preprogrammed in the system and should be available for recall whenever an operator needs help. This should include programmed instruction materials and Computer Assisted Instruction (CAI) for use by inexperienced operators.

- Restart procedures and checkpoints should be built into the system so they can be utilized during a transaction, should the operator make an error.

- The system should be designed with format aids. Preselected formats force an operator to work through the system in an orderly manner and discourage browsing. Preprinted forms may be useful so the operator does not have to remember which items must be entered into the system. Light pens might be utilized to make the selections from the preselected formats easier and less error-prone.

- Adequate work area (light, noise, temperature, etc.) is a must if operator errors are to be reduced to a minimum.

- Reasonably fast response time (2 to 4 seconds) is required if an operator is to utilize the system with peak efficiency. Sometimes a response of less than 2 seconds is not believable and one of more than 4 seconds may make the operator impatient and more error prone.

- To reduce operator boredom, graphics, color, flashing elements, varying brightness, varying character size, broken lines, and varying widths of lines should be intermixed.

- The terminal might have a split platen or screen to allow for a scratch pad work area. With this, the operator has a work area that does not interfere with a continuous form that may be in the other half of the platen. This not only increases throughput, it also reduces errors, because the operator can perform calculations before entering the actual data.

- Intelligent terminals can be utilized to edit and verify the data prior to transmission. In this way, many errors will be detected and corrected before transmission of the data to the central computer.

Efficiency and Redundancy

Earlier in this chapter, we used the term "efficiency." Now let us enlarge on the idea. One objective of a data communication network is to achieve the highest possible volume of accurate information through the system. The higher the volume, the greater the resulting systems efficiency and the lower the cost. Systems efficiency is affected by such characteristics of the lines as distortion, transmission speed, as well as by turnaround time on two-wire lines, the coding scheme utilized, the speed of the transmitting and receiving equipment, the error detection and control methodologies, and the mode of transmission. In this section, we will focus on the last two factors.

Transmission efficiency is defined as the fraction of the full theoretical capacity of the system that actually carries information. Thus efficiency is measured by:

$$\text{Efficiency} = \frac{\text{Information/Unit Time}}{\text{Capacity/Unit Time}}$$

As noted earlier in this chapter, a 7-bit USASCII code that uses the 8th bit for parity has an efficiency

$$E = \frac{7}{8} = 0.875 \times 100 = 87.5\%.$$

This efficiency refers to the code scheme only. Depending upon the transmission method, the overall efficiency will drop still further.

Redundancy is a concept that is complementary to efficiency. The term refers to superfluous or non-data-carrying bits added to the original data. Examples are parity bits, forward error-correction bits, and polynomial check bits on blocks of data. Redundancy may be calculated by the following relationship:

$$\text{Redundancy} = 1 - \text{Efficiency}.$$

For example, a 7-bit USASCII code is 0.125 redundant.

The above discussion centered on the efficiency and redundancy of codes. The system designer is more concerned with the efficiency of the entire system, including the combined coding and transmission efficiency (effective speed). In a transmission system, losses of efficiency are caused by the coding scheme and by the method used to transmit the data. For example, during synchronous transmission, an efficiency loss is caused by the extra parity bit and a small loss is caused by the synchronization characters transmitted before and after each block of data. In asynchronous or isochronous transmission, an efficiency loss is caused by the start and stop synchronization bits that are added to each character before its transmission. For example, consider an asychronous transmission scheme that sends an 8-bit byte preceded and followed by 1 start and 1 stop bit, respectively. Then the efficiency of the transmission scheme (as distinct from the code) is 0.8, because 10 bits are sent to convey 8 bits of data. Teletype, which uses one start and two stop bits has $E = \frac{8}{11} = 0.727$. The reader might want to refer again to Figure 2-9 to determine the bits of information and the total bits transmitted for several codes.

In order to determine the efficiency of the combined system elements, the system designer should multiply the efficiency of the coding scheme by the efficiency of the transmitting equipment:

$$\text{Efficiency (total system)} = \text{Efficiency (coding scheme)} \times \text{Efficiency (transmission equipment).}$$

Using the above example for USASCII coding, the efficiency of the total system would be:

$$E = 0.875 \times 0.80 = 0.70.$$

Assume that the system designer is going to utilize a system that transmits data at 4,800 bps. If the designer is going to use a USASCII code that has 7 data bits and 1 parity bit as well as 1 start bit and 1 stop bit in asynchronous transmission, then the total system efficiency is only 0.70; that is, $0.70 \times 4,800$ gives an effective throughput of 3,360 bps ($4,800 \times 0.70 = 3,360$).

This is an important factor in designing a system, because the designer may have to achieve a certain minimum information transfer rate. Depending on the throughput efficiency of the total system, the designer may utilize different coding structures to improve the coding scheme efficiency and different modes of transmission to improve the transmission equipment efficiency.

In a practical sense, many characters are required to make a message and the loss of one character (or the insertion of a spurious character) may destroy the meaning of a transmitted message. Further, it is important to include source and destination data in a message so that the receiver will know that he is not receiving a misdirected message. Such provisions also cost transmission time. Finally, most transmission equipment requires the use of certain control characters to mark significant parts of the messages transmitted.

For example, assume that messages are being sent on a network of 50 teletype terminals using asynchronous transmission of USASCII code at 110 bps. The text character count ranges from 10 to 300 characters and averages 100. The messages have internal checking in two forms:

- A 3-digit number equal to the number of characters in the text is appended to each message
- Two 2-digit numbers identifying the sending and receiving terminals are also appended to each message.

Further, assume that special control characters will be inserted in the message at four points

- At the beginning
- Between the terminal numbers and the character count
- Between the character count and the text
- At the end of the text.

Then, for this example

$$\text{Efficiency} = \frac{\text{average number of text characters}}{\text{average number of total message characters}} \times$$

$$\frac{\text{number of information bits/character}}{\text{number of transmitted bits/character}}$$

Efficiency = 100 / (4 control characters + 4 terminal ID characters + 3 text character count + 100 text characters)

$$\times \frac{7}{11} = \frac{100}{111} \times \frac{7}{11} = 0.573$$

The effective transmission speed at 110 bits per second is:

$$0.573 \times 110 = 63 \text{ bps.}$$

Blocking Factors

An additional consideration arises in synchronous transmission where an entire block of data is transmitted at one time. The tradeoff of effective speed versus the block length may become a problem to the system designer. A short block length creates an adverse ratio of transmission time to turnaround time because of excessive "overhead" when the sending device transmits a short block, the receiving device acknowledges the correct or incorrect receipt of that block, and the sending device sends another short block. A long block length increases the probability of error (since errors occur in bursts) and may require extensive retransmission time. Thus, the long block length increases the probability of errors and the length of the block increases the retransmission time.

To get an idea of the efficiency of block transmission, calculate the time to send one block and for the receiver to acknowledge it, using a half-duplex line. Assume the block contains N data characters and C control characters, i.e., the block length is $N + C$. If the transmission time for one character is T_C, then the time to transmit all characters will be $T_C (N + C)$. To acknowledge the receipt of the block will require one terminal delay, T_T, and one line turnaround time, T_L. After the acknowledgement, another T_T and T_L will have to elapse in order to prepare to send again. Therefore, the total time per block is

$$T_C (N + C) + 2 (T_T + T_L).$$

Now calculate the information transmission rate. If there are B bits per character, then the information transmission rate, R_I, is:

$$R_I = \frac{B \cdot N}{T_C(N + C) + 2(T_T + T_L)} \quad \text{(in bits/unit time)}$$

Then the efficiency can be obtained by dividing the information transmission rate by the theoretical transmission rate, R_T, (in bits per unit time).

$$E = \frac{R_I}{R_T} = \frac{B \cdot N}{R_T[T_C(N + C) + 2(T_T + T_L)]}$$

For example, assume

B = 8 bits/character
N = 80 character block length
C = 20 control characters
R_T = 4,800 bps
T_C = 1.7 milliseconds (ms) (8 bits/character ÷ 4,800 bps)
T_T = 60 ms
T_L = 150 ms

$$E = \frac{8 \cdot 80}{4,800\,[0.0017(80 + 20) + 2(0.060 + 0.150)]}$$

$$E = 0.226$$

Increasing the block size by doubling N to 160 increases the efficiency to 0.367. Doubling it again, to 320, yields an efficiency of 0.534. The above formula applies to half-duplex lines only, because it includes line turnarounds which are unnecessary in full-duplex transmission or if the echo suppressors are disabled.

In the case of full duplex, replace the line turnaround time, T_L, in the equation with a new, smaller quantity T_S, the time for the modem to synchronize with the incoming transmission. Depending on line speed and modem type, T_S lies in the range of 5 to 50 milliseconds.

Assume we are to drive a remote line printer from a 9,600 bps, full-duplex line. What printing rate can be achieved, using 120 character information records? The other facts are assumed to be:

B = 8 bits per character
N = 120 characters
C = 12 characters
T_C = 0.83 ms (= 8 bits/character ÷ 9,600 bps)

$T_T = 60$ ms
$T_S = 20$ ms

Total time per block $= T_C(N + C) + 2(T_T + T_S)$
$= 0.83(120 + 12) + 2(60 + 20)$
$= 110 + 160 = 270$ ms/block

or $\dfrac{1.000}{0.270} = 3.70$ blocks (printed lines) per second

or $60 \times 3.70 = 222$ printed lines per minute.

All the above formulas omit one potentially important factor—errors. If 1% of all the blocks have at least one error, then the information rate will drop by $1\% + (1\%)^2 + (1\%)^3 + \ldots$ (The extra terms represent errors occurring when *re*-transmitting, *re-re*transmitting, etc.) What is the likelihood of 1% of the blocks having an error? Suppose the bit error rate is 1 in 10^5; then with 8-bit characters, the character error rate is 1 in 12,500.* If x is the block length, $(N + C)$, at the 1% error rate then:

$x = 0.01 \times$ character error rate
$x = 0.01 \times 12,500$
$x = 125 = N + C$ for 1% block error rate.

This does not appear to be very serious. However, with increasing block length, the error rate causes the situation to deteriorate at an increasing rate because of the increased likelihood of repeated retransmissions. (It may also create queuing problems—see Chapter 8.) The system designer must be alert to these potential problems, and should control their occurrence by applying knowledge and analytical ability to their solution.

Questions—Chapter 5

As you can see, error control involves some of the most mathematical and intellectually stimulating topics in data communications. Coding and related techniques touch upon many familiar activities. The questions in this chapter are again of four kinds: True or False, Fill-In, Multiple Choice, and Short Answer. Within the last group, Question 2 deals with a technique that

---

*$(1 \times 10^5/8 = 12,500)$

may be used on one or more of the credit cards you are carrying. Question 5 shows the form of error protection used on all standard computer magnetic tapes. Use these questions to test your understanding; error control is a vital part of data communications.

True or False

1. Simple character checking schemes are very effective in detecting virtually all errors in data communication systems.

2. Shorter message blocks have the advantages of having a lower probability of error and a greater efficiency of transmission than longer blocks.

3. Intermodulation noise is a special type of echo noise.

4. Attentuation involves sudden changes in the level of power.

5. The 4-of-8 code is an example of a constant ratio code.

6. Forward error correction always requires 100% redundancy.

7. The Hamming code requires 1 parity bit per character.

8. If the coding scheme is 87.5% efficient, then the transmission will be 12.5% redundant.

9. The efficiency of the combined elements of a system is the sum of the individual efficiencies of the elements in the system.

10. The time for a modem to synchronize with the incoming transmission is typically longer than for a line turnaround.

Fill-in

1. Errors usually come in ——— in data communications.

2. ——— ——— occurs when one line picks up some of the signal that is traveling down another line.

3. USASCII is an example of a code using ——— checking.

4. ——— is the process of transforming a message that is in plain text to cipher text.

5.   A reasonable response time of approximately —— to
     —— seconds is required if an operator is to utilize a data
     communication system with peak efficiency.

6.   CAI stands for —— —— —— .

7.   In the formula $R_I = (B \cdot N)/[T_C(N + C) + 2(T_T + T_L)]$, R_I
     stands for —— —— .

8.   In the above formula, B is the —— —— —— .

9.   In the above formula, N is the —— —— —— —— .

10. In the above formula, C is the —— —— —— —— .

Multiple Choice

1.   When transmitting over the public switch networks, the
     error rate may be increased by:

     a) Transmitting during periods of high traffic
     b) Transmitting data at lower speed
     c) Utilizing line conditioning
     d) Using private leased lines
     e) None of the above

2.   The type of noise caused by the thermal agitation of elec-
     trons is:

     a) Impulse noise
     b) Jitter
     c) Line outages
     d) White noise
     e) None of the above

3.   The main source of errors in data communications is:

     a) Impulse noise
     b) Jitter
     c) Line outages
     d) White noise
     e) All are equally responsible for errors

4.   Cross talk decreases with which of the following:

     a) Decreased distances between wires
     b) Decreased signal strength
     c) Higher frequency signals
     d) Increased communication distances
     e) None of the above

5.   An equalizer compensates for which of the following:

   a) Attentuation
   b) Delay distortion
   c) Intermodulation noise
   d) A and B above
   e) None of the above

6.   Telegram communication best illustrates which of the following:

   a) Error detection with automatic correction
   b) Error detection with retransmission
   c) Ignoring errors
   d) Loop or echo checking
   e) All of the above

7.   Every message is transmitted at least twice using:

   a) Error detection with automatic correction
   b) Error detection with retransmission
   c) Ignoring errors
   d) Loop or echo checking
   e) None of the above

8.   If a system is going to use a code with 7 data bits, 1 parity bit, 1 start bit, and 2 stop bits per character, the combined system efficiency is:

   a) over 0.99
   b) 0.875
   c) about 0.727
   d) 0.70
   e) less than 0.70

Short Answer

1.   Why does the 4-of-8 code not use a parity bit?

2.   A frequently used form of checking of decimal numbers is the "modulo nine" check. A check digit is appended to each numerical quantity. The digit is the remainder, upon division by nine, of the sum of the digits in the original number. For example, if the original number is 73842, then the sum of the digits is 24, $(7 + 3 + 8 + 4 + 2)$. The check digit calculation is: $24/9 = 2 +$ Remainder of 6; therefore, the check digit is 6. The number with the check digit appended would then be 738426. Assume that during transmission there is an error that changes the number

738426 to the number 739426. Let us check the received number (739426). The sum of the digits is 25, (7+3+9+4+2); therefore, 25/9 = 2 + Remainder of 7. Since 7 does not equal 6, the number has been detected in error. Experiment with this method, finding out under what circumstances it works and when it fails. If you are particularly industrious, look up "casting out 9's" in a mathematical dictionary or investigate this property in a book on number theory.

3. How efficient is the method above if we are transmitting 8-digit numbers? 12-digit numbers?

4. For the following array of code groups calculate the parity bits using even parity for the rows and odd parity for the columns.

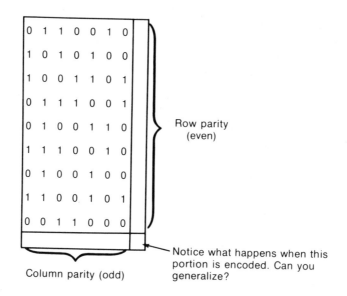

Row parity (even)

Column parity (odd)

Notice what happens when this portion is encoded. Can you generalize?

5. Using the format of Figure 5-4, encode the data character 1011.

6. Find an on-line terminal in use in some business situation (point of sale, banking, etc.) and examine it for features that relate to error control. List and describe the features that you found.

7. Calculate the efficiency of transmission in the following situation:

- Block length = 1,000 data characters plus 20 control characters.
- Character length = 10 bits.
- Transmission rate = 4800 bps.
- Terminal delay = 150 ms.
- Turn around time = 125 ms.

8. In the above situation, suppose the bit error rate is 2 in 10^6, what is the likelihood of an error in a block?

9. Do errors in data communications normally appear in bursts or are they distributed evenly in time?

10. A signal suffers a loss of power as it travels from the transmitting device to the receiving device. What is this called?

11. What is the most used approach to error control?

6

Data Communication Software

Up to now, we have not touched on software.
Nonetheless, software plays many roles in data
communication systems. An understanding of the
principles of systems organization is a necessary
prerequisite to specifying and designing data
communications-related software. Software of this type
includes programs to perform functions common to all
applications, and to control the network and terminal
equipment and the application programs themselves.
For best results, the organization and basic logic of
these programs must follow certain principles of
design, and a carefully planned software testing
program must be conducted.

A Systems Context

Up to this point, we have been dealing largely with facts about
the physical and logical components of data communication systems. One of the major functions of software is to act as the
"mortar" that holds these components together and fills in the
gaps where adjoining components fail to fit smoothly. Because of
this pervasive quality, it is most profitable to look at software in
an overall systems context.

Over the years, many approaches to formulating such a
context have been made, often with substantial success. In 1975,

IBM announced Systems Network Architecture (SNA) as its approach. While it is not ideal, we believe SNA will grow to be the common form of terminology and discourse for communications-oriented systems because of the wide distribution of IBM equipment and software, and the high number of professionals trained in those products.

The concept of SNA provides a generalized, hierarchical, and modular description of the data communication environment. As such, it will tend to organize thinking about systems structure and will result in implementations sharing certain underlying characteristics that have been found to be beneficial. In other words, we discuss SNA here as a way of *thinking* about data communication systems, rather than as a product of one manufacturer. At this point, our objective is to teach *principles*. Keep this viewpoint in mind as you read the following section.

Systems Network Architecture (SNA)

SNA describes an integrated structure that provides for all modes of data communications and upon which new data communication networks can be planned and implemented. SNA is built around four basic principles. First, SNA encompasses distributed functions in which many network responsibilities can be moved from the central computer to other network components, such as remote concentrators. Second, SNA describes paths between the end users (programs, devices, or operators) of the data communication network separately from the users themselves, thus allowing network configuration modifications or extensions without affecting the end user. Third, SNA uses the principle of device independence, which permits an application program to communicate with an input/output device without regard to any unique device requirements. This also allows application programs and communication equipment to be added or changed without affecting other elements of the communication network. Fourth, SNA uses standardized functions and protocols, both logical and physical, for the communication of information between any two points. This means that there can be *one* architecture for general purpose and industry terminals of many varieties, and *one* network protocol.

A data communication network built on SNA concepts can be considered to consist of:

- A central computer ("host node")

- A front-end communication processor ("intermediate node")

- Remote concentrators ("intermediate node")

- A variety of general purpose and industry oriented terminals ("terminal node" or "cluster node" if a group of terminals and a local controller are involved).

SNA portrays the communication system in terms of physical and logical entities. Conceptually, this involves a physical network and a logical network. The physical entities in the communication system are the nodes and links of Figure 6-1. The nodes are the distinct hardware components that are capable of performing information processing and network control. The host node is the central computer. The intermediate node is primarily concerned with the routing and transmission of data. An intermediate node to which a cluster or terminal node is attached is also known as a boundary node. The SNA intermediate node is a local or remote front-end communication processor under the

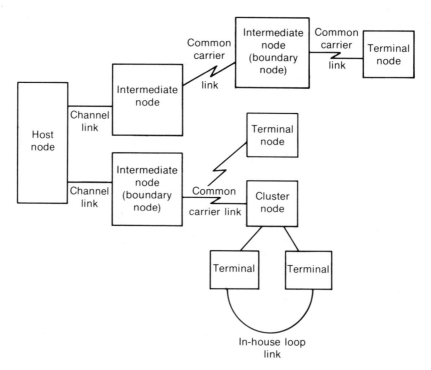

Figure 6-1: Systems Network Architecture Physical Network

control of a Network Control Program (NCP). The cluster node, which may also operate under program control, provides the information processing and network control capabilities for a group of local or remotely attached devices. The terminal node, which provides the lowest level of capability and intelligence in the network, is basically concerned with the input and output of information through terminal devices. Links comprise the media used for information transfer between the nodes. The connecting links include the channels between the central computer and the front-end communication processor, the common carrier facilities, and on-site loops. The SNA network is, from a physical point of view, a set of nodes interconnected by links.

The SNA logical network consists of three layers, the application layer, the functional management layer, and the transmission subsystem layer. Each node in the SNA physical network may contain any or all of these three layers. Figure 6-2 depicts the SNA logical network.

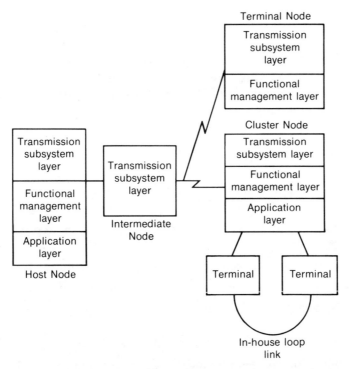

Figure 6-2: Systems Network Architecture Logical Network

The application layer consists of the user's application programs and is concerned only with the processing of information. In this concept, the application program is isolated from the idiosyncrasies and complexities of unique hardware device operations and data communications.

The functional management layer provides the interface to the communication system for the application programs. Its primary concern is the formatting of information transmitted and received in the communication network.

The transmission subsystem layer performs the routing, scheduling, and transmission functions associated with the movement of information between nodes in the communication system.

The segregation of functions assigned to software in these three layers is the key to simplicity, efficiency, and reliability in a data communication system. A designer who assigns duties in two or more layers to one program can anticipate trouble. To plan a communication system using SNA principles, the designer ideally should be concerned only with the application layer. The services of the functional management and transmission subsystem layers should be available to the user as standard manufacturer-supplied software and should be transparent to that user. If not, considerable extra effort lies ahead for the implementer. In this case he is faced with two unattractive alternatives. He can build the functions missing from the other layers into his application programs or into the manufacturer-supplied software. Both approaches are costly and difficult.

A link control concept called Synchronous Data Link Control (SDLC) is a part of the overall SNA concept. It is a discipline for the management of information transfer over data communication lines. The SDLC function includes the following control activities:

- Synchronizing, or getting the transmitter in step with the receiver
- Detecting and recovering transmission errors
- Controlling the sending and receiving between stations
- Reporting improper data link control procedures.

The SDLC procedures take each message and sandwich it into a frame for transmission. In the SDLC concept, the frame is the vehicle for every command and response, and for all information that is transmitted using SNA. Figure 6-3 depicts an SDLC frame. All messages are put into this frame format and transmitted from one node to another node. The error checking for each

Flag (start)	Device address	Control field	Message	Frame check sequence	Flag (end)

- Control field: sequence number of next frame we expect to receive, sequence number of next frame we expect to send, and the like.
- Frame check sequence: uses a cyclic algorithm

Figure 6-3:   SDLC Transmitting Frame

message is incorporated in the frame check sequence portion of the SDLC frame, whereas the control field checks for missing or duplicated frames.

Software Organization in Practice

The preceding discussion dealt with a somewhat abstract and taxonomical view of systems. Modern data communication software tends to follow SNA principles fairly well. The distinction between the application layer and the other layers is usually observed carefully, especially if the manufacturer-supplied communication software is sufficiently versatile and effective. The distinction between the functional management layer and the transmission subsystem layer is not clearly observed, however, so certain undesirable rigidities exist today in most prepackaged communication software.

Software in a data communication network can reside almost entirely in the central computer (host node), or part of it can be located in the front-end communication processor, in a remote concentrator (intermediate nodes), or in an intelligent terminal. Figure 6-4 depicts a typical software organization in a central processor and front-end communication processor. In this figure, the NCP resides in the front-end communication processor, although some of the communication input/output buffer operations reside in the central processor. Besides the communication input/output buffers, the host node contains:

- An operating system that coordinates the interactions between the application programs and the SNA access method
- The various application programs
- An access method control program that acts as the teleprocessing manager

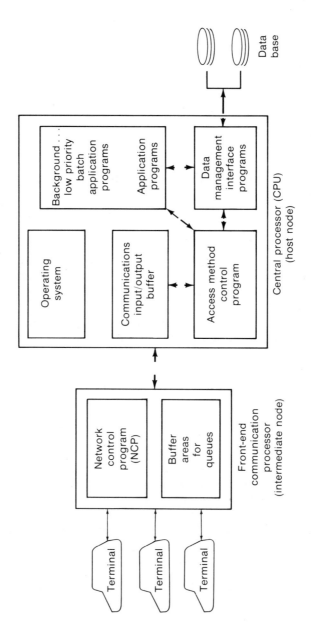

Figure 6-4: Typical Software Organization in the Host and Intermediate Nodes

- Background programs
- Data management interface programs.

Normally included in the functions of the NCP are:

- Activation and deactivation of the links and stations
- Establishment of logical connections over physical links
- Processing of network commands
- Detection and correction of line errors and polling
- Recording of network status.

As networks get larger the systems designer can initiate migration of the NCP functions outward into the network. Figure 6-5 depicts a situation where these programs have been moved to a remote concentrator.

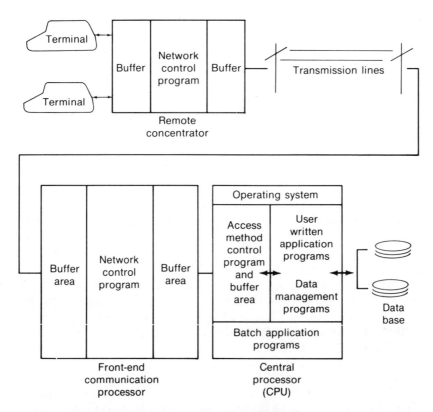

Figure 6-5: Alternative Configuration of Software Organization in a Distributed Data Communication Network

Examples of Communication Software

The most familiar and highly developed communication software packages are the access method control programs of the IBM 360/370 series, some of which have been in use since the late 1960s. They cover a range of capabilities, and to varying degrees, may include aspects of the functional management layer in addition to fulfilling the duties of the transmission subsystem layer software. There are four such packages, which are presented as examples of the range of products with which the designer can work:

- Basic telecommunication access method (BTAM)
- Queued telecommunication access method (QTAM)
- Telecommunication access method (TCAM)
- Virtual telecommunication access method (VTAM).

The Basic Telecommunication Access Method (BTAM) provides the basic functions needed for controlling data communication lines in IBM 360/370 systems. It supports asynchronous terminals, binary synchronous communications, and audio response units. BTAM is a set of basic modules that may be used to construct communication programs. It is recommended for use where there are 10 or less lines to support or when a specialized communication control program is required. BTAM requires a knowledge of the terminal's operation, link discipline, and a basic knowledge of programming. A BTAM user must write routines for the scheduling and allocation of facilities (the functional management layer). The basic flow control and data administration routines are also the responsibility of the BTAM user (additional parts of the functional management layer). It is the least sophisticated of the four data communication software programs listed above, but it does contribute the lowest system overhead. BTAM provides facilities for polling terminals, transmitting and receiving messages, detecting errors, automatically retransmitting erroneous messages, translating code, dialing and answering calls, logging transmission errors, allocating blocks of buffer storage (OS360/370 only), and performing on-line diagnostics to facilitate the testing of terminal equipment. BTAM resides in the central computer and is the interface between the front-end communication processor and the user written application programs.

The Queued Telecommunication Access Method (QTAM) is an extension of BTAM and includes all the BTAM facilities except that it does not support binary synchronous communications;

therefore, it supports only asynchronous terminals. QTAM provides a high level and flexible macro language for the control and processing of communication data, including message editing, queuing, routing, logging, and so on (the functional management layer). It can schedule and allocate facilities, poll terminals, perform error checking routines, reroute messages, cancel messages, and the like. QTAM is not utilized much any more and has largely been replaced with TCAM.

The Telecommunication Access Method (TCAM) replaces and extends the older QTAM. The most significant features of TCAM are those for network control and system recovery. An operator control facility is also provided for network supervision and modification. It supports asynchronous terminals, binary synchronous communications, and audio response units. TCAM performs all the functions of BTAM and QTAM, and handles the data communications in a system that utilizes a high degree of multiprogramming. Unlike the prior basic data communication software, TCAM has its own control program that takes charge and schedules the traffic handling operations. In some cases it can handle an incoming message by itself without passing it to an application program, for example routing a message to another terminal in a message switching system. TCAM also provides status reporting on terminals, lines, and queues. It has significant recovery and serviceability features to increase the security and the availability of the data communication system. The checkpoint and restart facilities are much more capable than those of QTAM. TCAM has prewritten routines for checkpointing, logging, date and time stamping, sequence numbering and checking, message interception and rerouting, error message transmission, and it supports a separate master terminal for the data communication system operator.

The Virtual Telecommunication Access Method (VTAM) is the data communication software package that complements IBM's advanced hardware and software, including the system 370. VTAM manages a network structured on SNA principles. It directs the transmission of data between the application programs in the central computer and the components of the data communication network. It operates with front-end communication processors. The basic services performed by VTAM include establishing, controlling, and terminating access between the application programs and the terminals. It moves data between application programs and terminals, permits application programs to share communication lines, communication controllers, and terminals. VTAM controls the configuration of the telecommunication network and permits the network to be monitored and altered in addition to performing all the basic functions of the other three data communication software packages.

In summary, BTAM is the least sophisticated access method and, to be effective, requires the most effort on the part of a user. QTAM has been replaced with TCAM. TCAM provides the facilities for a complete communication system. It supports a wide range of terminals, provides network control, significant recovery facilities, and now supports SNA. VTAM is the most advanced data communication software package and is intended to support the virtual computer systems using SNA. Figure 6-6 is a checklist of the functions that can be performed by this type of software.

Software Design Precepts

The data communication environment poses some unique problems for the program designer. The most basic of these is the lack of control over the time dimension. In conventional batch processing, the designer plans the program so it can refuse to deal with inputs until conditions are just right; if things go wrong, the designer simply stops the "clock" and causes the program to abort. This is a luxury not often available in the data communication world because inputs arrive at the computer with timing and in sequences beyond the control of the designer.

In Chapter 5, we said that communication errors are a fact of life. This fact also impacts heavily on the program designer. The schemes described in Chapter 5 deal mostly at the level of one to a few dozen characters. They are, of course, ineffective for catastrophic errors such as line breaks and interruptions lasting as long as one or more messages. The programs must accept the responsibility for prudent action in these situations. Finally, although it happens only infrequently, computer hardware fails, too. When it does, it produces its own set of problems for the program designer.

Basically, three factors make data communication programs different:

- Lack of control over input timing
- Communication errors
- Computer failures

- Message buffering
- Error checking
- Code conversion
- Link establishment/termination
- Message formatting
- Polling
- Management of task queues
- Timeout period checks
- Answering of incoming calls
- Dialing of outgoing calls
- Alerting of console operator to error conditions
- Priority classification assignments to terminals
- Message traffic statistics collection
- Message editing
- Processing of input-output requests
  from applications programs
- Logging
- Security/privacy
- Message compression

Figure 6-6: Checklist of Software Functions for a Data
Communication Network

Note that these problems impact primarily on the functional management and transmission subsystem layers. This implies the examination (and possible enhancement) of the properties of manufacturer-supplied software.

What can the program designer do to deal with these factors? The first step is to ensure that the software provides proper *message accountability*. Basically, message accountability is a record-keeping function that ensures no inputs or intended outputs "fall through the cracks" and that recovery from a communication or computer interruption can be accomplished with minimum damage to users.

For each incoming message this means:

- Logging as soon as received
- Time tagging
- Address checking
- Format, and where possible, content error checking
- Receipt acknowledgment after logging and checking
- Diagnosing and acting constructively on errors
- Maintaining statistics on errors

- "Turning off" terminals and lines that send excessive numbers of errors

For outgoing messages it is important to:

- Log at time of transmission
- Require acknowledgment, and act constructively if acknowledgment is not received
- Provide a priority scheme to ensure that outgoing overloads are worked off in a rational manner
- Test the integrity of lines and terminals and maintain statistics on results
- Provide a rational means of disposing of messages that can not be sent because of line or terminal errors.

The final step in ensuring message accountability combines inputs and outputs. The relationships that tie together inputs and outputs (e.g., "input message type A yields exactly one output message type B") must be put to work by coupling them to the input and output logs to provide a continuous statistical accounting of work in process, to detect any failures in obeying the relations, and to provide the proper basis for recovery processes invoked after failures.

The next step is to conduct a *failure mode analysis*. In this process, the software designer examines the consequences to message integrity of each possible failure in the system. These failures must be presumed to occur successively at each stage of program execution. Possible countermeasures are evaluated and selected for implementation by the software. This is an easy process to describe and a difficult and exhausting one to perform, but it is absolutely necessary if the resulting system is to have even minimal initial viability. It is likely that most of the software in the system will be devoted to dealing with the exceptions occasioned by communication, hardware, and people failures rather than to accomplish the "mainline" system functions. The failure mode analysis, therefore, can easily turn out to be the major component in the design task.

Finally, all the principles of good noncommunication program design also apply. The chief of these are:

- Modularity—Break the functional job up into small, "neat," functional modules, and match the program structure to the functional structure.
- Hierarchy—Recognize the hierarchy of the functional modules and mirror it in the module calling relationships

- Generality—Look for the truly "primitive" functions, generalize their definitions, and clearly identify their basic parameters so that modules can be defined to perform groups of similar functions, rather than proliferating specialized modules.

Software Testing Precepts

The same factors that make design of data communication software uniquely difficult also tend to make testing of that software more complex than in the batch environment. Because the timing and sequencing of "real world" inputs is not always predictable, it is difficult to build confidence that any testing procedure has sufficiently exercised the time relationships to uncover all time-dependent pathological behavior in the program under test. Similarly, in testing the response of the program to environmental factors such as communication errors and hardware failures, it is difficult to create a sufficient variety of these events to ensure that the program is thoroughly tested.

What can be done to ease these problems? There are three areas upon which the systems implementer should focus:

- Test planning
- Test execution
- Test documentation.

First, test planning is a function that is often forgotten until it is too late. Properly, the software test plan should be developed *as a part of the software functional specification*. This approach has several virtues. It helps ensure that:

- The test will focus on proving the *performance* of the software, rather than proving that the programmer's *concept* of the design matches the design itself.
- Test support facilities, such as computer time, special test data, communications, special test generation or data reduction programs, and the like, are identified early enough to plan their acquisition intelligently and efficiently.
- "Testability" of the software will be a design criterion. This will also improve the "diagnosability" of errors and the overall maintainability of the software.
- All parties in the systems development effort will know the criteria upon which the suitability of the software will be judged.

Second, test execution should be handled, if at all possible, by personnel different from those who developed the software. The objectivity introduced by an independent software test group will pay off in improved performance and software integrity. Users must participate in the conduct of the testing, if it is to be done effectively. Their involvement should increase progressively as the "bugs" are removed from the software and the detailed functional characteristics become more and more apparent. During this time, two things tend to happen. First, the user can offer immediate, first-hand pragmatic judgments about discrepancies between specifications and actual performance. Often such discrepancies can be removed in simple ways, e.g., using the application knowledge of the user. Second, the user is building knowledge of, and confidence in, the software.

Finally, test documentation is an often overlooked activity. "Coming events cast their shadows before them" is an adage that predates the software business by about 150 years,* but it has real importance in this current context. Almost all the failures in a software system, whether they are found during testing or weeks, months, or years after the software has been declared operational, "cast their shadows" during implementation and testing. Thorough, careful documentation of test planning, preparation, execution, and post-test analysis will provide the best possible groundwork from which to analyze the failures and prevent their recurrence. It is important to recognize that often more money is spent on "maintaining" real time programs (i.e., fixing failures not found during testing and upgrading functional capabilities) than was spent on developing the programs in the first place. Good documentation is the foundation of any effort to keep software maintenance costs in line.

Software Aids for Network Design

There is one other type of software of primary importance to the communication network designer. These are the software packages that evaluate network performance characteristics and automatically compute prices for leased line telecommunication networks using a wide variety of tariffs from the various common carriers. This type of software is employed primarily in the design

---

*Thomas Campbell (1777–1844) "Lochiel's Warning."

of networks that utilize voice grade lines because the tariff restrictions are too complicated. A typical description of a software package that optimizes the network design is as follows:

> The program computes prices for telecommunication networks consisting of leased lines and other devices or components having fixed monthly charges. It will automatically price point to point and multipoint leased line networks by accepting inputs for these links in the form of their end point vertical/horizontal coordinates. In the basic version of this software package, multipoint lines are priced by treating each constituent point to point segment as an individual input record. This package also has available, as an option, a single-circuit multipoint optimization feature. Here the end user need only input the terminal locations to be connected; the program automatically determines the least cost layout for the multipoint circuit. For each link to be priced, the user may select up to four AT&T interstate tariffs from the following list (Series 1001, 1002, 1003, 1005, 1006, 2001, 2006, 3002, 4002, 5700, 5800, 8000).
>
> Optionally available are MCI and Southern Pacific specialized carrier tariffs as well as 2400 BPS, 4800 BPS, 9600 BPS, and 56,000 BPS intercity DDS tariffs including both types of access lines for all DDS offerings. These options, which involve the evaluation of indirect routing and routing optimization by the program, are available in this software package.*

Questions—Chapter 6

Software integrity is crucial to the overall integrity of data communication systems. This chapter has given you an insight into this dependence and has set forth principles upon which this integrity is based. As in earlier chapters, the questions are of four kinds: True and False, Fill-In, Multiple Choice, and Short Answer. Short answer questions 2, 3, and 4 are designed to put your understanding of these principles to work in situations that are realistic aspects of the data communication system designer's job. The remaining short answer questions test your knowledge in other ways.

Network Pricing, Analysis, and Inventory Control Program by D. M. V. Telecommunications Corporation, 2975 Hickory Lane, Ann Arbor, Michigan 48104.

True or False

1. The SNA network is, from a physical point of view, a set of nodes interconnected by links.

2. Ideally, a designer should attempt to integrate his application programs and modifications to manufacturer-supplied software into one program as part of the effort to simplify overall system design.

3. NCP functions are virtually always exclusively concentrated in the central computer for larger systems.

4. The designer of a data communication program has less control over the arrival of inputs than in conventional batch processing systems.

5. Failure mode analysis is usually a major component in the software design task of a communication system.

6. Testing of communication software tends to be more complex than program testing in conventional batch processing systems.

7. Testability of the software should be a design criterion.

8. Software packages that evaluate network performance characteristics and automatically compute prices for leased line telecommunication networks are of importance to the communication designer.

Fill-in

1. SNA stands for —— —— ——.

2. The —— —— layer performs routing, scheduling, and transmission functions in the communication system.

3. The —— —— layer provides the interface to the communication system for the applications programs.

4. Using the SDLC concept, the —— is the vehicle for every command, every response, and all information that is transmitted using SNA.

5. NCP stands for —— —— ——

6. —— —— is a record-keeping function that ensures no inputs or intended outputs are lost and that recovery from

a communication or computer interruption can be accomplished with minimum damage to users.

7.   The software —— —— should be developed as a part of the software functional specification.

8.   —— —— should be handled by personnel different from those who develop the software.

9.   Good —— is the foundation of any effort to keep software maintenance costs down.

10. The —— can offer immediate, first-hand pragmatic judgments about discrepancies between specifications and actual performance.

Multiple Choice

1.   Which of the following are basic principles of SNA?

a) Device independence
b) Distributed functions
c) Standard functions and protocols
d) A and C above
e) All the above

2.   An intermediate node in a data communication network built on SNA concepts could be which of the following?

a) Central computer
b) Front-end communication processor
c) Remote concentrator
d) A and B above
e) B and C above

3.   The SNA logical network consists of which of the following:

a) Applications layer
b) Functional management layer
c) Transmission subsystem layer
d) A and C above
e) All the above

4.   The host node would normally contain:

a) Applications programs
b) Concentrators
c) Terminals

d) A and B above

e) B and C above

5. Which of the following provides the basic functions needed for controlling data communication lines in IBM 360/370 systems?

a) BTAM

b) QTAM

c) TCAM

d) VTAM

e) CTAM

6. Which of the following supports only asynchronous terminals?

a) BTAM

b) QTAM

c) TCAM

d) VTAM

e) CTAM

Short Answer

1. In the section on SNA, the four basic principles of SNA are listed. Discuss these principles, concentrating on the advantages that they provide.

2. In the section, "Software Design Precepts," "acting constructively on errors" is cited as important for handling incoming messages. Assume a banking situation in which a withdrawal message is being sent from a terminal to the central computer. Describe a "constructive action" for each of the following unusual conditions detected at the computer:

- Account number is garbled

- Requested withdrawal exceeds account balance by a small amount such as $100

- Requested withdrawal exceeds account balance by a very large amount such as $10,000

- The incoming transaction is unreadable.

3. Also in the Software Design Precepts section, it is suggested that both incoming and outgoing messages be logged. Explain how, in a store and forward message switching sys-

tem, these logs could be used to recover from a computer failure.

4. Propose a set of procedures for a store and forward message switching system to meet the requirements in the Software Design Precepts section. Provide a rational means of disposing of messages. Assume there are three priorities of messages: delivery as soon as possible, delivery in two hours, and delivery in 12 hours. Assume terminals can be repaired in 6 hours and lines can be repaired in 1 hour if they fail. Assume the messages are English text and their average length is 20 words.

5. What are the three basic factors that make data communication programs different from batch programs?

6. What is the first step to ensure that the software provides proper integrity of the messages being transmitted?

7

Common Carriers and Tariffs

The preceding six chapters have concentrated on
technical factors. This chapter focuses on business and
market factors, particularly costs. The systems
designer must be aware of communication line costs
as well as the technical characteristics of the
hardware that will be utilized. These costs can affect the design
and structure of a data communication
network more than any other single
factor. The systems designer must learn to
compare the costs and benefits of the data
communication network, and management must
evaluate these costs and benefits in terms of their
probable effect on the organization's goals and
objectives. This chapter introduces the common
carriers, the communication services offered, tariffs,
and examples of services and hardware costs.

Common Carriers

The concept of the common carrier is to provide transport for a
fee. Common carriage, a field with a long history, was exemplified
in its early stages by the obligation of ferryboat operators to serve
the general public at reasonable prices. Before the advent of
government regulation, common carriers were in a position to

charge unreasonable rates for the services that they offered the public and were under no obligation to provide regular service.

The electromagnetic telegraph for transmitting intelligence over electrical wires was invented by Samuel F. B. Morse in 1832. In 1843, Congress appropriated $30,000 to build an experimental telegraph line from Washington, D.C., to Baltimore. Early telegraph services were generally provided by railroad companies, who strung wires along their rights of way. Wire telegraph played an important part in the development of the West, with extensive transmissions by telegraph beginning in the early 1850s. The cost of a telegram in those days was as high as $20.

The Post Roads Act of 1866 authorized the Postmaster General to fix rates annually for government telegrams. In 1887, Congress gave the Interstate Commerce Commission (ICC) authority to require telegraph companies to interconnect their lines for more extended public service. At this time, there were over 50 small telegraph companies, none of which could send telegrams to another company.

On March 10, 1876, Alexander Graham Bell transmitted the first complete message heard over wire. Before this, all messages were sent using a telegraph key and Morse Code, which is composed of groups of dots and dashes (short and long pulses). The first regular telephone line was installed from Boston to Salem, Massachusetts, in 1877. By the end of 1880, there were almost 50,000 telephones in the United States. The dial telephone was invented by Almon B. Strowger of Kansas City in approximately 1889 and the first dial exchange (end office or central office) was installed in LaPorte, Indiana, in 1892. Prior to this dial exchange telephones were leased in pairs, with the subscriber having to put up his own lines to connect with another telephone subscriber.

Government regulation of the financial practices of common carriers began with the Mann-Elkins Act of 1910 which authorized the ICC to establish uniform systems of financial accounting records for telegraph and telephone common carriers. This act required that telephone and telegraph companies file monthly and annual financial reports with the ICC. Basically, the controlling forces over telephone and telegraph companies were the Post Office Department (this is still true today in most European countries), the ICC, and the Department of State, with much overlapping authority and redundancy.

In 1933, President Roosevelt appointed a committee to study the control of common carriers. The committee concluded that communication services should be regulated by a single body. It recommended the establishment of a new government agency that would regulate all interstate and foreign communication by wire, radio, telephone, and telegraph. In February 1934, the President sent a message to Congress urging the creation of the Federal Communications Commission (FCC). The resulting Communications Act, which created the FCC, was signed by the President in June 1934. This act was designed to make available rapid and efficient wire and radio communication services, with adequate facilities upon which messages could be transmitted nationwide and worldwide at reasonable charges.

The FCC is an independent agency that regulates, in the public interest, interstate and foreign communications by wire, radio, TV, and cable. It has jurisdiction in all 50 states, Guam, Puerto Rico, and the Virgin Islands. The Commission has wide regulatory powers, both to compel common carriers to conform to broad objectives of the Communications Act and to make any necessary inspections and investigations of common carriers. The FCC also has additional regulatory jurisdiction under provisions of the Communications Satellite Act of 1962. The FCC does not regulate Federal Government radio operation, however.

Every common carrier that engages in either interstate (between states) communication or international communication falls under the jurisdiction of the FCC. Common carriers that are solely intrastate (within one state) are not subject to FCC jurisdiction; instead, they come under the authority of their home state communication regulatory commissions.

For a common carrier to have its communication facilities approved by the FCC, it must file basic information documents with the FCC giving details of its intended service, the charges for that service, classifications, regulations, and the like. These documents are called tariffs and they form the basis of the contract between the common carrier and the user of the common carrier's service. Sometimes the tariffs of the giants, like the American Telephone and Telegraph Company, become standards that other companies use. Some of the current common carriers are:

Interstate Services

American Satellite Corporation
20030 Century Boulevard
Germantown, Maryland 20767

American Telephone & Telegraph Company, Long Lines
32 Avenue of the Americas
New York, New York 10015

General Telephone & Electronics
One Stamford Forum
Stamford, Connecticut 06904

Graphnet Systems, Inc.
99 W. Sheffield Avenue
Englewood, New Jersey 07631

MCI Communications Corporation
1150 17th Street N.W.
Washington, D.C. 20036

Nebraska Consolidated Communications Corporation
3240 South 10th Street
Lincoln, Nebraska 68502

RCA Global Communications, Inc.
60 Broad Street
New York, New York 10004

Southern Pacific Communications Company
1 Adrian Court
Burlingame, California 94010

Telenet Communications Corporation
1666 K Street N.W.
Washington, D.C. 20006

United States Transmission Systems, Inc.
Two Broadway
New York, New York 10004

Western Tele-Communications, Inc.
54 Denver Technological Center
Denver, Colorado 80201

Western Union Telegraph Company
One Lake Street
Upper Saddle River, N.J. 07458

International Services

American Telephone and Telegraph Company, Long Lines
32 Avenue of the Americas
New York, New York 10013

ITT World Communications, Inc.
67 Broad Street
New York, New York 10004

RCA Global Communications, Inc.
60 Broad Street
New York, New York 10004

TRT Telecommunications Corp.
1747 Pennsylvania Avenue N.W.
Washington, D.C. 20006

Western Union International, Inc.
26 Broadway
New York, New York 10004

In summary, a communications common carrier is a company that sells communication services to the public (individuals, partnerships, corporations, and so on). These common carriers fall under the jurisdiction of the FCC if the services they sell are either interstate or international in their connections. The intrastate common carriers are subject only to the regulatory commissions of the individual states within which they operate.

Communication Services

The American Telephone and Telegraph Company (AT&T) is the largest of the telephone and telegraph common carriers. It consists of about 20 separate, but closely interdependent companies. Two other large companies that offer a broad spectrum of data communication facilities are General Telephone and Western Union. There are other relatively large telephone companies, such as United Utilities of Kansas, Continental Telephone Company, Central Telephone Company, in addition to almost 1600 other small and very small telephone companies. These telephone companies, along with the specialized common carriers such as Southern Pacific Communications, Western Telecommunications, MCI, Graphnet Systems, RCA, American Satellite, and others, offer various data communication services. Some of these firms offer a broad range of services, while others offer only specialized services, such as communication lines between key cities, specialized forms of data transmission, or satellite circuits furnished exclusively to common carriers for resale by them.

On any given link of a network, the transmission speed is usually determined by both the type of line chosen and the particular modems used at the ends of the line. In order to determine which type of communication line is required, the designer must first have a general understanding of the types of communication services offered. Figure 7-1 depicts the most common types of communication services that are available today. Some very specialized types of services have been omitted, but the average system will use one of the communication services shown.

The communication services in this figure may be offered by more than one common carrier. For example, some 1600 telephone companies offer Direct Distance Dialing, and a few common carriers offer metered service. The term "WATS," however,

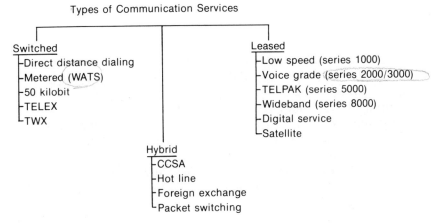

Figure 7-1:   Some Common Types of Communication
Services

is reserved for a specific metered service offered by AT&T. The service offered by most common carriers is voice grade service (series 2000/3000). The rest of this chapter will be devoted to discussing the five types of switched services, the four types of hybrid services, and the six types of leased services shown in Figure 7-1.

All the costs listed in this chapter are realistic in that they prevailed in 1977. They are rounded to the nearest whole dollar, wherever appropriate, and they are adequate for network design problems. Current costs, which change frequently, can be found in the FCC tariff publications or in *The Guide to Communications Services.**

Switched Services

Switched services allow the subscriber to use a dial telephone to make a connection to some other subscriber who also has a dial telephone. Switched services include the following basic offerings:

- Direct distance dialing
- Metered (WATS)
- 50 kilobit

---

*This can be obtained from the Center for Communications Management, Inc., P.O. Box 324, Ramsey, New Jersey 07446.

- TELEX
- TWX

Direct distance dialing (DDD) is nothing more than utilizing for data communications the dial telephone that most people have in their homes. In DDD, the user calls another station (terminal or computer), thus making a connection between his terminal and the distant station. In this way, the data communication link is established by using the regular switched telephone system described in Chapter 4.

The data communication user pays the same rate as does an individual who uses the telephone as a voice communication

	Sat	Sun	Mon	Tues	Wed	Thurs	Fri
8:00 AM to 5:00 PM			Day rate				
5:00 PM to 11:00 PM		Evening rate					
11:00 PM to 8:00 AM	Weekend and night rate						

Mileage Between Stations	Day		Evening		Night & Weekend	
	Initial 1 Minute	Additional 1 Minute	Initial 1 Minute	Additional 1 Minute	Initial 1 Minute	Additional 1 Minute
1−10	$ .19	$ .08	$ .1235	$ .052	$ .076	$ .032
11−16	.23	.11	.1495	.0712	.092	.044
17−22	.27	.13	.1755	.0845	.108	.052
23−30	.31	.17	.2015	.1105	.124	.068
31−40	.35	.20	.2275	.13	.14	.08
41−55	.39	.24	.2535	.156	.156	.096
56−70	.41	.26	.2665	.169	.164	.104
71−124	.43	.28	.2795	.182	.172	.112
125−196	.44	.29	.286	.1885	.176	.116
197−292	.46	.31	.299	.2015	.184	.124
293−430	.48	.33	.312	.2145	.192	.132
431−925	.50	.34	.325	.221	.20	.136
926−1910	.52	.36	.338	.234	.208	.144
1911−3000	.54	.38	.351	.247	.216	.152

Figure 7-2:   Direct Distance Dialing Rates (Station to Station)

device. Rates are based on the length of time that the user is on the line and the distance between the stations. Figure 7-2 can be used to determine the cost of utilizing this type of communication line for interstate service. Intrastate DDD service is slightly higher. For example, if the user calls between 8 A.M. and 5 P.M. on a Monday through Friday (see top of Figure 7-2) then the day rate is in effect. In the lower portion of Figure 7-2, notice that there are two columns for the day rate. The first column shows the initial minute of the call, and the second covers each additional minute of connect time. If the data communication link was between Los Angeles, California, and Tulsa, Oklahoma, the distance would be 1,266 miles. For this distance, the initial minute, for a day rate call, would cost $0.52 and each additional minute would cost $0.36. If this connection was made for three hours during the day, then the total cost of the call would be $64.96 (1 minute × $0.52 + 179 minutes × $0.36 = $64.96).

Metered (WATS)   services use voice grade lines for subscribers on the public dial telephone network. Wide Area Telecommunications Service (WATS) is a special bulk rate service offered by AT&T for directly dialed, station-to-station telephone calls. This service can be used for voice communications or for data transmission. The United States is divided into service areas and the geographical coverage from any one of these service areas is determined by the "band" of service to which the customer subscribes. Figure 7-3 depicts, as an example, the WATS bands

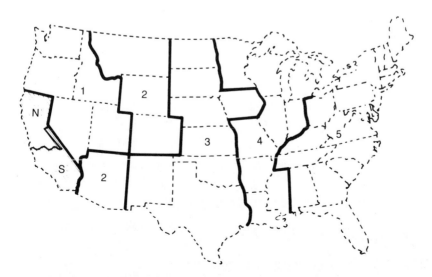

Figure 7-3:   WATS Bands from California

from either Northern or Southern California. There are, of course, other similar maps for the other service areas. Notice that **Band 1** generally affords interstate coverage to nearby states and **Band 5** provides nationwide coverage.

WATS service to a higher number band, such as Band 3, includes service to all the lower numbered bands. It is recommended that the reader contact a local telephone system office and request the WATS booklet that includes the prices and the local WATS map service area. As might be expected, the WATS bands from New York are the opposite of the WATS bands from California. For example, New York is in WATS Band 5 from California, and California is in WATS Band 5 from New York.

WATS is offered either on a measured time or a full business day basis. The basic charge for measured time WATS covers 10 hours of use per month, and full business day WATS covers 240 hours per month. Any time in excess of these figures is billed as an extra charge. Figure 7-4 shows the interstate WATS tariffs for WATS lines from California. Notice that the price differs, depending on whether the subscriber is located in Northern or Southern California, since Northern and Southern California are in separate service areas.

In data communication applications, where the average holding time per call is less than 1 minute, charges are based on the total number of calls at 1 minute per call. In other words, WATS rates are subject to a 1-minute minimum average call holding time. For example, a measured time WATS line allows 10 hours of

Service Area	Band 1	Band 2	Band 3	Band 4	Band 5
California (N) Full business day (240 hours/month)	$1,570.00* ($4.36)†	1,640.00 (4.55)	1,660.00 (4.61)	1,670.00 (4.63)	1,675.00 (4.65)
Measured time (10 hours/month)	$ 230.00 ($17.25)	238.00 (17.85)	242.00 (18.15)	244.00 (18.31)	245.00 (18.38)
California (S) Full business day (240 hours/month)	$1,610.00 ($4.47)	1,645.00 (4.56)	1,660.00 (4.61)	1,670.00 (4.63)	1,675.00 (4.65)
Measured time (10 hours/month)	$ 234.00 ($17.55)	239.00 (17.93)	242.00 (18.15)	244.00 (18.31)	245.00 (18.38)

*Dollars per month
†Costs in the parentheses represent the per hour charge in excess of either 240 or 10 hours per month, whichever the subscriber uses

Figure 7-4: Interstate WATS Tariffs

use per month; therefore, since 10 hours have 600 minutes, a maximum of 600 calls are allowed within the minimum service charge when leasing a measured time WATS line. A WATS line is arranged for either outward dialing or inward dialing. Inbound and outbound calls cannot be combined on a single WATS line.

In summary, the user of an interstate WATS line that leases, say, a Band 4 (outbound) WATS line has the privilege of calling, without additional charge, any location in Bands 1, 2, 3, and 4 for up to the number of hours prescribed. Thus, a Band 4 full business day WATS line from Northern California would cost $1,670 per month, plus $4.63 per hour for any hours in excess of 240 during the month.

The WATS bands shown in Figure 7-3 and 7-4 do not include calls within a state (in this case, California). In order to have either inbound or outbound WATS lines that can be utilized intrastate, separate intrastate WATS lines must be leased. For intrastate service, California offers either statewide service, Northern California service, or Southern California service.

Currently there is a tariff change proposed that will restructure WATS service. The proposed changes to the WATS service are contained in Tariff F.C.C. Number 259. The essence of these changes are, that instead of having the 48 contiguous United States divided into five bands, they will be divided into three bands which will be designated as A-1, A-2, and A-3. Band A-1 will generally afford interstate coverage to nearby states while band A-3 will provide nationwide coverage excluding Alaska and Hawaii. Also, there will be a new band A-4 which encompasses WATS service to Alaska and Hawaii. As before, WATS to a higher number band includes service to lower numbered bands.

The other major change will be in the costing structure. Instead of the more or less flat rates for 240-hour or 10-hour service, there will now only be service for the initial 10 or 30 hours with extra cost involved for time exceeding the 10 or 30 usage hours per month. For outward WATS the costs are broken down into categories of so many dollars for the initial 10 hours, the next 30 hours (11-40), the next 50 hours (41-90), the next 50 hours (91-140), the next 60 hours (141-200), and each additional hour beyond that. For example, the proposed cost of an outward WATS line from northern California to band A-3 would be $270.00 for the initial 10 hours, $17.00 per hour for the next 30 hours, $13.00 per hour for the next 50 hours, $10.00 per hour for the next 50 hours, $2.00 per hour for the next 60 hours, and $1.00 for each additional hour beyond 200; whereas, an inward WATS line to northern California (A-3) is proposed to cost $75.00 for the initial 30 hours, $13.00 per hour for the next 30 hours, $12.00 per hour for the next 40 hours, $10.00 per hour for the next 50 hours, $4.00 per hour for the next 70 hours, and $2.00 per hour for each additional hour beyond 220. Again, we recommend that the reader contact their local telephone company in order to obtain the latest WATS maps and rates as they emanate from their local geographic area.

50 Kilobit switched service is a measured-use, high-speed, data transmission service. The bandwidth of the channel provided on this dial-up service is 48,000 cycles per second (48 kHz). This is equivalent to 12 voice grade lines.

This service is especially useful for high-speed data communications and facsimile transmission. In addition to this basic facility, a voice grade channel is provided for voice communications, which is generally utilized for coordination between the sender and the receiver.

This service is utilized for transmission of data up to 50,000 bps. Currently, 50 kilobit is limited to service between most of the cities and suburbs located in and around Chicago, Los Angeles, New York City, San Francisco, and Washington, D.C. This service is especially useful for transmission of hard copy utilizing facsimile transmission devices. The cost of this service is shown in Figure 7-5. To calculate the cost, the system designer first determines the mileage between the two points that will be connected.

Mileage	Per Minute or Fraction Thereof
1–50	$0.50
51–150	0.80
151–300	1.25
301–600	1.75
601–1200	2.25
1201–2000	2.75
2001–over	3.25

Usage	Service Terminal*	
	Installation	Per Month
Facsimile device	$125.00	$275.00
Synchronous data (50,000 bps)	125.00	275.00

*A service terminal is the plug into which the user plugs a modem.

Figure 7-5: 50 Kilobit Service

For example, for a distance of 675 miles, the charges would be $2.25 per minute of connect time. Also, a service terminal is needed at each end of the line. The service terminal cost per month is also shown in Figure 7-5.

TELEX, or Teletypewriter Exchange Service, is a 66-word-per-minute Teletypewriter to Teletypewriter communication network offered within the United States and to certain points in Canada and Mexico. The subscriber uses Teletypewriters that are leased from the common carrier. Each subscriber has a private dial-up line and telephone number, as with a conventional telephone. The service includes a TELEX book that contains the TELEX number of the thousands of other subscribers to the service. The subscriber can call any other terminal on the network and transmit a message to that terminal. This is nothing more than an alternative to the voice telephone network but it offers the ability to transmit hard copy between subscribers. In this case, the common carrier is responsible for the terminal, its connections to the switched network, the lines over which the data are transmitted, the connections, and the terminal at the other end of the network. TELEX is offered by Western Union and, through facilities of the international carriers, to subscribers in Hawaii and some 135 countries abroad.

The rates for TELEX service are based on 12 service areas that function somewhat similarly to the WATS bands. For example, Rate Area 12 includes California and Nevada, while Rate Area 2 includes Florida, Georgia, South Carolina, and North Carolina. To calculate the cost for a call, one must consult a rate table to determine the charge per minute of connect time between rate areas. For example, from Area 2 to 12 it is $0.60 per minute. There is also a service terminal charge for each end of the line with TELEX service, as well as an initial installation charge. The cost of TELEX service is not shown because of its similarity to Teletypewriter Exchange Service (TWX) which is described next.

TWX, or Teletypewriter Exchange Service, is a Teletypewriter to Teletypewriter communication exchange between most points in the United States and Canada. TWX uses a 7-bit plus parity USASCII code and is available at two transmission speeds, 60 and 100 words per minute. Like TELEX, TWX is a service in which the common carrier is responsible for supplying a terminal, the connections to the line, and the line services. With TWX, however, the subscriber may provide the terminal instead of leasing it from the common carrier. The basic rates for TWX service are shown in Figure 7-6. As can be seen from this figure, the charge is

Between Points in the United States	
Mileage	Per Minute or Fraction Thereof
0-50	$0.20
51-110	0.25
111-185	0.30
186-280	0.35
281-400	0.40
401-550	0.45
551-750	0.50
751-1030	0.55
1031-1430	0.60
1431-2000	0.65
2001-over	0.70

Usage	Terminal	
	Installation	Per Month
60 wpm (model 28 ASR)	$50.00	$119.00
100 wpm (model 33 ASR)	50.00	79.00
100 wpm (model 35 ASR)	50.00	150.00
Customer interface unit	50.00	34.00

Figure 7-6:   TWX—Teletypewriter Exchange Service

calculated by determining the mileage between the two calling points and applying the cost per minute of connect time. The lower half of Figure 7-6 shows the cost of utilizing either a 60-word-per-minute terminal, a 100-word-per-minute terminal, or the customer's own terminal. For example, a 60-word-per-minute Model 28 automatic send receive (ASR) terminal costs $50.00 for installation, plus $119.00 per month. On the other hand, if the customer provides the terminal, the cost is $50.00 for installation and $34.00 per month for the plug to which the customer's terminal is connected (customer interface unit).

Hybrid Services

Hybrid services include some characteristics of switched services and some characteristics of leased services. They also may be a special type of service unto themselves. The hybrid services include:

- CCSA
- Hot Line

- Foreign exchange
- Packet switching

CCSA (Common Control Switching Arrangement) is a private long-distance dialed network. This is a special service consisting of intercity telephone circuits priced under a special tariff rate or under the private line leased rate, whichever is lower. This service can be purchased by large organizations to interconnect several business operations and thus save on their telephone costs. An example of this is the United States Government's Federal Telecommunications System (FTS). FTS is a CCSA network that is leased from AT&T and is utilized by government employees for calls inside the United States. Each agency utilizing this system must be physically connected to it. Many thousands of PBXs are connected to this network. Lines from these many thousands of switchboards go into consolidated switchboards that are connected to the FTS system. Approximately 250 consolidated switchboards in the United States are interconnected by the CCSA network (FTS) that is leased from AT&T. This CCSA network arrangement is utilized by the Federal Government for voice communications and data transmission within the United States. It has been estimated that this common control switching arrangement saves the Federal Government more than $125 million per year over the cost of using regular telephone service and paying the standard public telephone rates.

Hot Line service is a metered use, private line telephone service between two points only ("point-to-point"). Hot Line is offered by Western Union and is not compatible with the normal telephone toll network. However, it can be terminated at a customer's PBX.

Hot Line directly connects two telephones in distant cities. When either of the two receivers are lifted, the telephone rings at the other end of the connection. This type of service is only available to selected cities and is not widely used. The cost of using this service ranges from $50.00 to $55.00 per month service charge, and from $0.16 to $0.65 per minute of connect time. Selected examples of the cost of Hot Line service are shown in Figure 7-7.

Foreign exchange service (FX) allows a subscriber to call another exchange area via the dial-up telephone network without incurring any charge other than that of a local call. For example, if an organization is located in the suburbs of a major city and it has

Selected Locations				Monthly Service Charge	Per Minute
Between	New York City	and	San Francisco	$50.00	$0.65
			Washington	55.00	0.16
			St. Louis	55.00	0.30
			Pittsburgh	55.00	0.18
			Miami	50.00	0.50
			Kansas City	55.00	0.35
	Chicago		New York City	55.00	0.27
			Detroit	55.00	0.16
			Minneapolis	55.00	0.19
			Denver	50.00	0.50
			Dallas	55.00	0.28
			San Francisco	50.00	0.60
			Cleveland	55.00	0.18

Figure 7-7: Hot Line Service

many calls to or from the downtown metropolitan area, it might lease an FX service so its telephones would, in effect, be connected to the telephone company's end office in the downtown metropolitan area. In other words, an FX line is a line that runs from your telephone instrument to the telephone company's end office in another area. In this way, you have all the free dialing privileges of any telephone subscriber in that area. In fact, when you pick up the telephone instrument you do not receive a dial tone from your own city, but from the distant city's end office. By listing your number in that distant city's phone book, people may call you from there, using what is to them a local number, and reach you over the FX line for which you are paying. The cost of FX service is the same as the cost for a voice grade private leased line. In other words, the cost of FX service is the same as leasing a telephone line between your premises and the distant telephone end office to which you want to be connected. These costs are covered later in this chapter in the section "Voice Grade (Series 2000/3000)."

Packet switching services (sometimes called "value-added networks" or VAN) provide a mixture of data communications and data processing services. The packet switching vendor leases lines from an existing common carrier and "adds value" to these leased lines by supplementing the lines with packet switching capabilities. Basically, packet switched data service is a switched data communication service that offers transmission speeds from 50 bps up to 56,000 bps and beyond in the future. The difference in the speed of transmission and the protocols used between the various stations is compensated for by the network minicomputers. Generally, data is segmented into 128-character (1024-bit) blocks called packets. Automatic error detection and correction of transmitted packets is a standard feature of packet switching data service.

In packet switching, the user's data are transmitted in packets over the packet switched network. In other words, the user does not have a network. The packets are transmitted over the vendor's communication network. There is no guarantee that all the packets of a specific message will be transmitted over the same line. This means that various packets of the same message may be routed alternately. This is transparent to the user (i.e., it does not affect the user) and it offers further security of the user's data during transmission.

Figure 7-8 shows an example of a packet switched network configuration. The minicomputers A through E are owned by the packet switching network and the lines connecting them are leased from a common carrier. The packet switching vendor then sells its services to various users. The two user terminals in Figure 7-8 represent organizations that have decided to use a packet switched network rather than to develop their own data communication network. When User Terminal 1 transmits to User Terminal 2, the messages may go from minicomputer A to E to C, or by any other route that is available when the packet is to

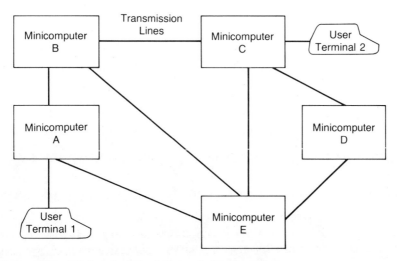

Figure 7-8: Packet Switched Network Configuration

be transmitted. In other words, the minicomputers determine which lines are free and transmit along that path. There is increased security potential for a user's data because the path upon which the data packet will travel is unknown until the moment of transmission.

Usually, the total monthly cost for packet switched data service network utilization is composed of a monthly usage charge per packet, a network access arrangement (port) charge, and the network interface equipment charge. A typical monthly usage charge is $0.60 per 1,000 packets or fraction thereof (1,000 packets × 128 characters* = 128,000 characters). Packets are not distance sensitive for pricing. The network access arrangement charges are for the access line to the local telephone end office (if dial access is desired) or the network channel termination charge (if a private leased line is desired). Either way, the user pays the network access arrangement costs.

The cost of a network access arangement for a leased private line is shown in Figure 7-9. For the network access arrangement costs, the standard communication common carrier rates apply for private line channels between the central office of the packet switched vendor and the customer's premises. In other words, these costs do not include the local loop between the customer's premises and the central office used by the packet switched network vendor. Also, at this time, this service is only offered to a select number of cities which are designated as high density, medium density, and low density. Selected examples of high-density areas are Boston, Chicago, Dallas, Detroit, Los Angeles, New York, Philadelphia, San Francisco, and Washington, D.C. Selected examples of medium-density areas are Atlanta, Baltimore, Cleveland, Denver, Houston, Minneapolis, Newark, Pittsburgh, and St. Louis. Selected examples of low-density areas are Albany, Buffalo, Columbus, Hartford, Indianapolis, Kansas City, and New Haven.

The last part of the total monthly charge for packet switched services is the network interface equipment charge. One example of this interface equipment is an access controller that provides up to 32 asynchronous access ports, each of which may operate at fixed transmission speeds of 50 bps, 75 bps, 110 bps, 134.5 bps, 150 bps, 300 bps, 600 bps and 1,200 bps. Therefore, the access controller is a concentrator that provides 32 variable speed access

---

*The user should try to tailor the message block size to the packet size to avoid sending messages with 50 characters and being charged for a 128 character packet.

Network Access Arrangement (Leased Line Port)*				
Transmission Speed (bps)	Installation	High Density (hourly)	Medium Density (hourly)	Low Density (hourly)
50–300	$250.00	$ 75.00	$110.00	$145.00
600	300.00	125.00	185.00	N/A
1200	350.00	150.00	250.00	300.00
2400–56000	400.00	200.00	N/A	N/A

Network Interface Equipment		
Item	Installation	Monthly
Basic equipment cost	$600.00	$400.00
Each access port		
50 bps – 300 bps	25.00	25.00
600 bps	40.00	40.00
1200 bps	50.00	50.00

*Standard common carrier rates apply for private leased lines between the end office used by the packet switching firm and the user's premises.

Figure 7-9: Packet Switched Service Cost

ports on the user organization's side and synchronous speeds of 2,400 bps to 56,000 bps on the other side. One use for this access controller might be to concentrate up to 32 multiple low-speed network ports to one higher speed data stream going to a central computer. A more likely use might be to concentrate up to 32 asynchronous input terminals into one high-speed data transmission output that would interface with the packet switched network for transmission to another location.

Leased Services

Leased private line services are communication services that are dedicated to the user. These private line facilities are offered as point-to-point communication lines or multipoint communication lines. This section will define and discuss the types of leased services that were listed in Figure 7-1:

- Low speed (Series 1000)
- Voice grade (Series 2000/3000)
- TELPAK (Series 5000)
- Wideband (Series 8000)

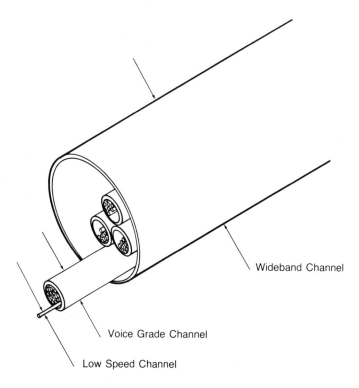

Figure 7-10:   Leased Services Analogy

- Digital service
- Satellite

Basically, **TELPAK**, wideband, digital service, and satellite service fall into the wideband channel category, even though digital and satellite service also offer transmission speeds in the voice grade category. Figure 7-10 shows an analogy, using water pipes, of the difference between low speed, voice grade, and wideband channels.

Low speed (Series 1000) private leased lines are used for low-speed signaling (e.g., burglar alarms) and Teletypewriter data transmission. These lines are used basically for transmission speeds up to 150 bps. The series 1000 lines are divided into five offerings, as shown below:

Type 1001: 1-30 bps channel for remote metering, supervisory control and miscellaneous signaling.

Type 1002: 1-55 bps channel for Teletypewriter, data or remote supervisory control and miscellaneous signaling purposes.

Type 1003: 1-55 bps channel for remote operation or radiotelegraph.

Type 1005: 1-75 bps channel for Teletypewriter, Teletypesetter, data or remote metering, and the like.

Type 1006: 1-150 bps channel for Teletypewriter, foreign exchange Teletypewriter, data or remote metering, and the like.

Interexchange Mileage Rates (per mile per month)*				
	First 100	Next 150	Next 250	Next 500	Each Additional Mile
Type 1001	$1.25	$1.00	$0.60	$0.40	$0.25
Type 1002	1.25	1.00	0.60	0.40	0.25
Type 1003	1.25	1.00	0.60	0.40	0.25
Type 1005	1.25	1.00	0.60	0.40	0.25
Type 1006	1.55	1.25	0.80	0.50	0.30

*Half/full duplex are the same price

Figure 7-11: IXC Per Mile Per Month Charge for Low-Speed Lines

Figure 7-11 shows the costs of each of the five types of low-speed lines. The Interexchange channel (IXC) mileage distance is calculated using special vertical and horizontal coordinates; this will be explained in Chapter 8.

For example, Figure 7-12 shows that for an IXC mileage of 325 miles for a Type 1003 line, the per month charge would be $320.00. If the user decided to interconnect to City C with a terminal (multidrop), then the charge would be $380.00. This example shows that every time a terminal is dropped off in a city, the user must go back to the beginning of the IXC per mile per month charge table (i.e., start back at the rate for the first 100 miles and work through the table).

There are other charges to be added to a leased low-speed line, including the service terminals for all five types of service (Types 1001 to 1006) and a channel terminal charge. Figure 7-13

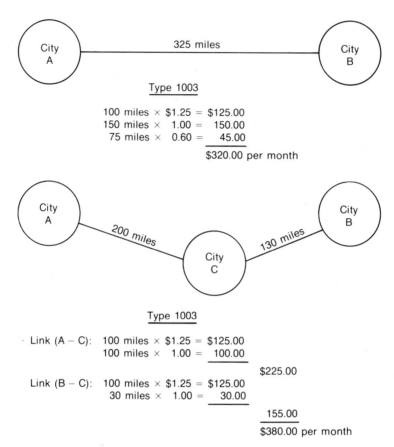

Figure 7-12:   Low-Speed Line Charges

depicts these various charges. Service terminals are charged at one price for the first service terminal in a telephone exchange (end office) and at a reduced price for each additional service terminal in that same telephone exchange (end office). For example, if a user has 10 telephone lines coming onto the premises, all from the same local telephone company end office, the user will be charged $40.00 for the first Type 1001 (HDX) service terminal and only $25.00 for each additional Type 1001 (HDX) service terminal. Figure 7-13 also depicts the channel terminal charges for Series 1000 lines. Figure 7-14 depicts an example of the cost of a Type 1006 (FDX) line between two major cities.

Voice grade (Series 2000/3000).   Because the dial-up telephone network uses voice grade lines for voice communication between

Service Terminals (per service terminal per month)
For the first station in exchange

	Installation	Half Duplex	Full Duplex
Type 1001	$52.55	$40.00	$44.00
Type 1002	52.55	40.00	44.00
Type 1003	52.55	40.00	44.00
Type 1005	52.55	40.00	44.00
Type 1006	52.55	60.00	66.00

Each additional station, same exchange

	Installation	Half Duplex	Full Duplex
Type 1001	$52.55	$25.00	$27.50
Type 1002	52.55	25.00	27.50
Type 1003	52.55	25.00	27.50
Type 1005	52.55	25.00	27.50
Type 1006	52.55	40.00	44.00

Channel Terminals: A monthly charge of $30.00 applies for each channel terminal of each two-point section except when the termination takes place at an international boundary point. This applies at each end of each circuit segment.

Figure 7-13: Service Terminals and Channel Terminals for Low-Speed Lines

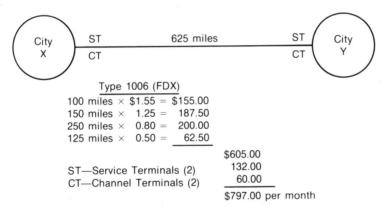

Figure 7-14: Type 1006 (FDX) Line Costs

subscribers, voice grade lines are the most popular form of leased line as well as being the most prevalent form of communication service. These services are offered in Series 2000, which includes voice transmission, mobile radio telephones, and foreign exchange service, and in Series 3000 voice grade lines, which are primarily for use in data communications. The rates for Series 2000 or 3000 are identical. The rates are based on the IXC mileage and the charges are on a per mile per month basis. The

Figure 7-15: Interexchange Mileage Charges for Voice Grade Lines

Schedule I (applies between a pair of Category A rate centers)		
Mileage	Charge	
1	$ 51.00	
2– 14	$ 51.00	+ $1.80 for each mile over 1 mile
15	$ 76.20	
16– 24	$ 76.20	+ $1.50 for each mile over 15 miles
25	$ 91.20	
26– 39	$ 91.20	+ $1.12 for each mile over 25 miles
40	$108.00	
41– 59	$108.00	+ $1.12 for each mile over 40 miles
60	$130.40	
61– 79	$130.40	+ $1.12 for each mile over 60 miles
80	$152.80	
81– 99	$152.80	+ $1.12 for each mile over 80 miles
100	$175.20	
101–999	$175.20	+ $0.66 for each mile over 100 miles
1000	$769.20	
over 1000	$769.20	+ $0.40 for each mile over 1000 miles

Where one rate center is an international boundary point, charge is as determined above, minus $25.00.

Schedule II (applies between a pair of rate centers where one rate center is in Category A and the other rate center of the same pair of rate centers is in Category B)		
Mileage	Charge	
1	$ 52.00	
2– 14	$ 52.00	+ $3.30 for each mile over 1 mile
15	$ 98.20	
16– 24	$ 98.20	+ $3.10 for each mile over 15 miles
25	$129.20	
26– 39	$129.20	+ $2.00 for each mile over 25 miles
40	$159.20	
41– 59	$159.20	+ $1.35 for each mile over 40 miles
60	$186.20	
61– 79	$186.20	+ $1.35 for each mile over 60 miles
80	$213.20	
81– 99	$213.20	+ $1.35 for each mile over 80 miles
100	$240.20	
101–999	$240.20	+ $0.66 for each mile over 100 miles
1000	$834.20	
over 1000	$834.20	+ $0.40 for each mile over 1000 miles

Where one rate center is an international boundary point, charge is as determined above, minus $25.00.

Schedule III (applies between a pair of Category B rate centers)			
Mileage	Charge		
1	$ 53.00		
2– 14	$ 53.00	+ $4.40 for each mile over	1 mile
15	$114.60		
16– 24	$114.60	+ $3.80 for each mile over	15 miles
25	$152.60		
26– 39	$152.60	+ $2.80 for each mile over	25 miles
40	$194.60		
41– 59	$194.60	+ $2.10 for each mile over	40 miles
60	$236.60		
61– 79	$236.60	+ $1.60 for each mile over	60 miles
80	$268.60		
81– 99	$268.60	+ $1.35 for each mile over	80 miles
100	$295.60		
101–999	$295.60	+ $0.68 for each mile over	100 miles
1000	$907.60		
over 1000	$907.60	+ $0.40 for each mile over	1000 miles

Where one rate center is an international boundary point, charge is as determined above, minus $25.00.

method of calculating the charges for a voice grade line varies from the method that was just described for low-speed lines. The basic charges for voice grade lines are shown in Figure 7-15. The tariff, Multischedule Private Line service (MPL), contains two rate elements, i.e., one for station terminals and one for IXC mileage (the costs are the same for either FDX or HDX service).

The charges for the IXC mileage are based on the category of city (Category A or B) as shown in Figure 7-15. There are 380 cities in Category A. All other cities are designated as Category B. Schedule I covers charges between Category A cities; Schedule II covers rates between Category A and Category B cities; and Schedule III covers rates between Category B cities.

Under Schedule I, monthly rates range from $51.00 for a 1-mile circuit plus $1.80 for each additional mile to $769.20 for a 1,000-mile circuit with $0.40 for each additional mile. Comparative rates for Schedule II are $52.00, with $3.30 for each additional mile, to $834.20 plus $0.40 for each additional mile. For Schedule III, the respective rates are $53.00 plus $4.40 per mile to $907.60 plus $0.40 for each additional mile. Under the MPL tariffs, the telephone company will price out the minimum-cost networks for users of Series 2000 and 3000 circuits.

In addition to the IXC mileage charges there are also station terminal (same as service terminal) charges for voice grade lines.

Station Terminal	Installation	Per Month
First station in exchange	$54.00	$25.00
Each additional station in same exchange		
Same user premises as first station	$54.00	$25.00
Different user premises	$54.00	$25.00

Figure 7-16:   Station Terminal Charges for Voice Grade Lines (Series 2000/3000)

The station terminal charges are levied by the telephone company to drop a line (local loop) to the customer's premises and provide an RS232 plug. The station terminal charges for voice grade lines are shown in Figure 7-16 .

The rate centers for Category A cities are listed below. All other cities are Category B rate centers.

CATEGORY A RATE CENTERS

Alabama
  Anniston
  Birmingham
  Decatur
  Huntsville
  Mobile
  Montgomery
  Troy

Arizona
  Flagstaff
  Phoenix
  Prescott
  Tucson
  Yuma

Arkansas
  Fayetteville
  Forrest City
  Hot Springs
  Little Rock
  Pine Bluff
  Searcy

California
  Anaheim
  Bakersfield
  Chico
  Eureka
  Fresno
  Hayward

Los Angeles
Oakland
Redwood City
Sacramento
Salinas
San Bernardino
San Diego
San Francisco
San Jose
San Luis Obispo
Santa Rosa
Stockton
Sunnyvale
Ukiah
Van Nuys

Colorado
  Colorado Springs
  Denver
  Fort Collins
  Fort Morgan
  Glenwood Springs
  Grand Junction
  Greeley

Connecticut
  Bridgeport
  Hartford
  New Haven
  New London
  Stamford

Delaware
  Wilmington

District of Columbia
  Washington Zone 1

Florida
  Chipley
  Clearwater
  Cocoa
  Crestview
  Daytona Beach
  Fort Lauderdale
  Fort Myers
  Fort Pierce
  Fort Walton Beach
  Gainesville
  Jacksonville
  Key West
  Lake City
  Miami
  Ocala
  Orlando
  Panama City
  Pensacola
  St. Petersburg
  Sarasota
  Tallahassee
  Tampa
  West Palm Beach
  Winter Garden
  Winter Haven

*International boundary points for Canada and Mexico.

Georgia
  Albany
  Atlanta
  Augusta
  Columbus
  Conyers
  Fitzgerald
  Macon
  Thomasville
  Waycross

Idaho
  Boise
  Pocatello
  Twin Falls

Illinois
  Centralia
  Champaign-Urbana
  Chicago
  Collinsville
  De Kalb
  Hinsdale
  Marion
  Mattoon
  Newark
  Northbrook
  Peoria
  Rockford
  Rock Island
  Springfield

Indiana
  Bloomington
  Evansville
  Fort Wayne
  Indianapolis
  Muncie
  South Bend
  Terre Haute

Iowa
  Boone
  Burlington
  Cedar Rapids
  Davenport
  Des Moines
  Iowa City
  Sioux City
  Waterloo

Kansas
  Dodge City
  Kansas City
  Manhattan
  Salina
  Topeka
  Wichita

Kentucky
  Danville
  Louisville
  Madisonville
  Winchester

Louisiana
  Alexandria
  Baton Rouge
  Lafayette
  Lake Charles
  Monroe
  New Orleans
  Shreveport

Maine
  Calais Int. Bdry*
  Portland

Maryland
  Baltimore
  Washington Zone 3

Massachusetts
  Boston
  Brockton
  Cambridge
  Fall River
  Framingham
  Lawrence
  Springfield
  Worchester

Michigan
  Detroit
  Flint
  Grand Rapids
  Houghton
  Iron Mountain
  Jackson
  Kalamazoo
  Lansing
  Petoskey
  Plymouth
  Pontiac
  Port Huron
    Int. Bdry*
  Saginaw
  Traverse City

Minnesota
  Duluth
  Minneapolis
  St. Paul
  Wadena

Mississippi
  Biloxi

Columbus
Greenville
Greenwood
Gulfport
Hattiesburg
Jackson
Laurel
McComb
Meridian

Missouri
  Cape Giradeau
  Joplin
  Kansas City
  St. Joseph
  St. Louis
  Sikeston
  Springfield

Montana
  Billings
  Helena
  West Sweetgrass
    Int. Bdry*

Nebraska
  Grand Island
  Omaha
  Sidney

Nevada
  Carson City
  Las Vegas
  Reno

New Hampshire
  Concord
  Dover
  Littleton
  Manchester
  Nashua

New Jersey
  Atlantic City
  Camden
  Hackensack
  Morristown
  Newark
  New Brunswick
  Trenton

New Mexico
  Albuquerque
  Las Cruces
  Roswell
  Santa Fe

New York
  Albany

---

*International boundary points for Canada and Mexico.

Binghamton
Buffalo
Buffalo Peace-
  bridge Int.
  Bdry*
Huntington
Mooers Forks
  Int. Bdry*
Nassau Zone 5
New York City
Poughkeepsie
Rochester
Syracuse
Westchester
  Zone 8

North Carolina

Asheville
Charlotte
Fayetteville
Greensboro
Greenville
Laurinburg
New Bern
Raleigh
Rocky Mount
Wilmington
Winston-Salem

North Dakota

Bismarck
Casselton
Crosby Inter-
  national -
  Crossing
  Int. Bdry*
Dickinson
Fargo
Grand Forks
Neche Int.
  Bdry*

Ohio

Akron
Canton
Cincinnati
Cleveland
Columbus
Dayton
Toledo
Youngstown

Oklahoma

Muskogee
Oklahoma City
Tulsa

Oregon

Medford
Pendelton
Portland

Pennsylvania

Allentown
  (Lehigh Co.)
Altoona
Harrisburg
Philadelphia
Philadelphia -
  Suburban
  Zone 26
Philadelphia -
  Suburban
  Zone 33
Pittsburgh
Reading
Scranton
Williamsport

Rhode Island

Providence

South Carolina

Charleston
Columbia
Florence
Greenville
Orangeburg
Spartanburg

South Dakota

Huron
Sioux Falls

Tennessee

Chattanooga
Clarksville
Jackson
Johnson City
Knoxville
Memphis
Nashville

Texas

Abilene
Amarillo
Austin
Beaumont
Corpus Christi
Dallas
El Paso
Fort Worth
Harlingen

Houston
Laredo
Laredo
  Int. Bdry*
Longview
Lubbock
Midland
San Angelo
San Antonio
Sweetwater
Waco

Utah

Logan
Ogden
Provo
Salt Lake
  City

Vermont

Burlington
White River
  Junction

Virginia

Blacksburg
Leesburg
Lynchburg
Newport News
  Zone 1
Norfolk Zone 2
Petersburg
Richmond
Roanoke
Washington
  Zone 8

Washington

Kennewick
North Bend
Port Angeles
  Int. Bdry*
Seattle
Spokane
Yakima

West Virginia

Beckley
Charleston
  Zone 1
Clarksburg
Fairmont
Huntington
  Zone 1
Morgantown

Parkersburg	Eau Claire	Wyoming
Wheeling	Green Bay	Casper
Zone 1	La Crosse	Cheyenne
	Madison	
Wisconsin	Milwaukee	
Appleton	Racine	
Dodgeville	Stevens Point	

Figure 7-17 shows the standard calculation for Series 2000/3000 charges and Figure 7-18 shows the same for a multidrop line.

As a check on the IXC mileage costs only, the "quick-rate calculator"* shown in Figure 7-19 will be useful. For line segments of 100 miles or less, Part 1 is used. For line segments of greater than 100 miles, the sum of Parts 1 and 2 will yield the appropriate rate as follows:

Example—683 miles for Schedule II cities:
```
600 miles from Part 1 ......................................$570.20
83 miles from Part 2  ....................................  54.78

   Total IXC charge  ....................................$624.98
```

Voice grade channels have a bandwidth of 3,000 cycles per second and, with current technology, this allows transmission speeds up to 9,600 bps. Voice grade lines may be conditioned when transmitting in the higher speed ranges (see conditioning in

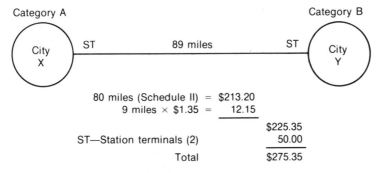

```
Category A                                    Category B

   ┌────┐                                      ┌────┐
   │City│  ST        89 miles        ST        │City│
   │ X  │                                      │ Y  │
   └────┘                                      └────┘

        80 miles (Schedule II)  =  $213.20
            9 miles × $1.35     =    12.15
                                   ─────────
                                    $225.35
        ST—Station terminals (2)     50.00
                                   ─────────
                        Total       $275.35
```

Figure 7-17: Voice Grade Line Charges Using MPL Tariffs Between a Category A and a Category B City

---

*International boundary points for Canada and Mexico

*The quick-rate calculator was developed by the Center for Communications Management, Inc., P.O. Box 324, Ramsey, New Jersey 07446.

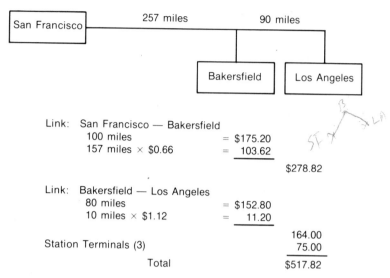

Figure 7-18: Multidrop Line Costs (Series 2000/3000)
Between Category A Cities

the glossary). Conditioning results in fewer errors during transmission and less delay distortion, cross talk, and echoes. Line conditioning can be performed only on private leased voice grade lines. Line conditioning costs range from $6.00 per month up to $50.00 per month. For example, depending on the level of conditioning desired, a point-to-point line conditioning charge ranges from $6.00 per month to $40.00 per month. For multipoint arrangements, the monthly line conditioning charges range from $11.00 per month to $50.00 per month.

TELPAK (Series 5000) lines are used for wideband applications. TELPAK is a group pricing arrangement of voice grade private line services. TELPAK was introduced in 1961 for both business organizations and government use. The two basic types of TELPAK lines are (1) Type 5700, a 240 kilohertz (kHz) carrier with a maximum equivalent channel capacity of 60 voice grade lines, and (2) Type 5800, a 1 megahertz (MHz) carrier (1,000 kHz) with a maximum equivalent channel capacity of 240 voice grade lines. The service is a pricing device and, unless otherwise specified and paid for, it will consist of separate voice grade lines usable up to the maximum number specified. Thus, the voice grade channels in TELPAK may be widebanded by means of special terminal

SERIES 2000/3000 QUICK-RATE CALCULATOR

PART 1				PART 2			
MILES	I	II	III	MILES	A $.68	B $.66	C $.40
0001	0051.00	0052.00	0053.00	0001	0000.68	0000.66	0000.40
0002	0052.80	0055.30	0057.40	0002	0001.36	0001.32	0000.80
0003	0054.60	0058.60	0061.80	0003	0002.04	0001.98	0001.20
0004	0056.40	0061.90	0066.20	0004	0002.72	0002.64	0001.60
0005	0058.20	0065.20	0070.60	0005	0003.40	0003.30	0002.00
0006	0060.00	0068.50	0075.00	0006	0004.08	0003.96	0002.40
0007	0061.80	0071.80	0079.40	0007	0004.76	0004.62	0002.80
0008	0063.60	0075.10	0083.80	0008	0005.44	0005.28	0003.20
0009	0065.40	0078.40	0088.20	0009	0006.12	0005.94	0003.60
0010	0067.20	0081.70	0092.60	0010	0006.80	0006.60	0004.00
0011	0069.00	0085.00	0097.00	0011	0007.48	0007.26	0004.40
0012	0070.80	0088.30	0101.40	0012	0008.16	0007.92	0004.80
0013	0072.60	0091.60	0105.80	0013	0008.84	0008.58	0005.20
0014	0074.40	0094.90	0110.20	0014	0009.52	0009.24	0005.60
0015	0076.20	0098.20	0114.60	0015	0010.20	0009.90	0006.00
0016	0077.70	0101.30	0118.40	0016	0010.88	0010.56	0006.40
0017	0079.20	0104.40	0122.20	0017	0011.56	0011.22	0006.80
0018	0080.70	0107.50	0126.00	0018	0012.24	0011.88	0007.20
0019	0082.20	0110.60	0129.80	0019	0012.92	0012.54	0007.60
0020	0083.70	0113.70	0133.60	0020	0013.60	0013.20	0008.00
0021	0085.20	0116.80	0137.40	0021	0014.28	0013.86	0008.40
0022	0086.70	0119.90	0141.20	0022	0014.96	0014.52	0008.80
0023	0088.20	0123.00	0145.00	0023	0015.64	0015.18	0009.20
0024	0089.70	0126.10	0148.80	0024	0016.32	0015.84	0009.60
0025	0091.20	0129.20	0152.60	0025	0017.00	0016.50	0010.00
0026	0092.32	0131.20	0155.40	0026	0017.68	0017.16	0010.40
0027	0093.44	0133.20	0158.20	0027	0018.36	0017.82	0010.80
0028	0094.56	0135.20	0161.00	0028	0019.04	0018.48	0011.20
0029	0095.68	0137.20	0163.80	0029	0019.72	0019.14	0011.60
0030	0096.80	0139.20	0166.60	0030	0020.40	0019.80	0012.00
0031	0097.92	0141.20	0169.40	0031	0021.08	0020.46	0012.40
0032	0099.04	0143.20	0172.20	0032	0021.76	0021.12	0012.80
0033	0100.16	0145.20	0175.00	0033	0022.44	0021.78	0013.20
0034	0101.28	0147.20	0177.80	0034	0023.12	0022.44	0013.60
0035	0102.40	0149.20	0180.60	0035	0023.80	0023.10	0014.00
0036	0103.52	0151.20	0183.40	0036	0024.48	0023.76	0014.40
0037	0104.64	0153.20	0186.20	0037	0025.16	0024.42	0014.80
0038	0105.76	0155.20	0189.00	0038	0025.84	0025.08	0015.20
0039	0106.88	0157.20	0191.80	0039	0026.52	0025.74	0015.60
0040	0108.00	0159.20	0194.60	0040	0027.20	0026.40	0016.00
0041	0109.12	0160.55	0196.70	0041	0027.88	0027.06	0016.40
0042	0110.24	0161.90	0198.80	0042	0028.56	0027.72	0016.80
0043	0111.36	0163.25	0200.90	0043	0029.24	0028.38	0017.20
0044	0112.48	0164.60	0203.00	0044	0029.92	0029.04	0017.60
0045	0113.60	0165.95	0205.10	0045	0030.60	0029.70	0018.00
0046	0114.72	0167.30	0207.20	0046	0031.28	0030.36	0018.40
0047	0115.84	0168.65	0209.30	0047	0031.96	0031.02	0018.80
0048	0116.96	0170.00	0211.40	0048	0032.64	0031.68	0019.20
0049	0118.08	0171.35	0213.50	0049	0033.32	0032.34	0019.60
0050	0119.20	0172.70	0215.60	0050	0034.00	0033.00	0020.00
0051	0120.32	0174.05	0217.70	0051	0034.68	0033.66	0020.40
0052	0121.44	0175.40	0219.80	0052	0035.36	0034.32	0020.80
0053	0122.56	0176.75	0221.90	0053	0036.04	0034.98	0021.20
0054	0123.68	0178.10	0224.00	0054	0036.72	0035.64	0021.60
0055	0124.80	0179.45	0226.10	0055	0037.40	0036.30	0022.00
0056	0125.92	0180.80	0228.20	0056	0038.08	0036.96	0022.40
0057	0127.04	0182.15	0230.30	0057	0038.76	0037.62	0022.80
0058	0128.16	0183.50	0232.40	0058	0039.44	0038.28	0023.20
0059	0129.28	0184.85	0234.50	0059	0040.12	0038.94	0023.60
0060	0130.40	0186.20	0236.60	0060	0040.80	0039.60	0024.00
0061	0131.52	0187.55	0238.20	0061	0041.48	0040.26	0024.40
0062	0132.64	0188.90	0239.80	0062	0042.16	0040.92	0024.80
0063	0133.76	0190.25	0241.40	0063	0042.84	0041.58	0025.20
0064	0134.88	0191.60	0243.00	0064	0043.52	0042.24	0025.60
0065	0136.00	0192.95	0244.60	0065	0044.20	0042.90	0026.00
0066	0137.12	0194.30	0246.20	0066	0044.88	0043.56	0026.40
0067	0138.24	0195.65	0247.80	0067	0045.56	0044.22	0026.80
0068	0139.36	0197.00	0249.40	0068	0046.24	0044.88	0027.20
0069	0140.48	0198.35	0251.00	0069	0046.92	0045.54	0027.60
0070	0141.60	0199.70	0252.60	0070	0047.60	0046.20	0028.00
0071	0142.72	0201.05	0254.20	0071	0048.28	0046.86	0028.40
0072	0143.84	0202.40	0255.80	0072	0048.96	0047.52	0028.80
0073	0144.96	0203.75	0257.40	0073	0049.64	0048.18	0029.20
0074	0146.08	0205.10	0259.00	0074	0050.32	0048.84	0029.60
0075	0147.20	0206.45	0260.60	0075	0051.00	0049.50	0030.00
0076	0148.32	0207.80	0262.20	0076	0051.68	0050.16	0030.40
0077	0149.44	0209.15	0263.80	0077	0052.36	0050.82	0030.80
0078	0150.56	0210.50	0265.40	0078	0053.04	0051.48	0031.20
0079	0151.68	0211.85	0267.00	0079	0053.72	0052.14	0031.60
0080	0152.80	0213.20	0268.60	0080	0054.40	0052.80	0032.00

Figure 7-19: Series 2000/3000 Quick Rate Calculator

SERIES 2000/3000 QUICK-RATE CALCULATOR (CONT)

PART 1				PART 2		
MILES	I	II	III	MILES	A $.68	B $.66	C $.40
0081	0153.92	0214.55	0269.95	0081	0055.08	0053.46	0032.40
0082	0155.04	0215.90	0271.30	0082	0055.76	0054.12	0032.80
0083	0156.16	0217.25	0272.65	0083	0056.44	0054.78	0033.20
0084	0157.28	0218.60	0274.00	0084	0057.12	0055.44	0033.60
0085	0158.40	0219.95	0275.35	0085	0057.80	0056.10	0034.00
0086	0159.52	0221.30	0276.70	0086	0058.48	0056.76	0034.40
0087	0160.64	0222.65	0278.05	0087	0059.16	0057.42	0034.80
0088	0161.76	0224.00	0279.40	0088	0059.84	0058.08	0035.20
0089	0162.88	0225.35	0280.75	0089	0060.52	0058.74	0035.60
0090	0164.00	0226.70	0282.10	0090	0061.20	0059.40	0036.00
0091	0165.12	0228.05	0283.45	0091	0061.88	0060.06	0036.40
0092	0166.24	0229.40	0284.80	0092	0062.56	0060.72	0036.80
0093	0167.36	0230.75	0286.15	0093	0063.24	0061.38	0037.20
0094	0168.48	0232.10	0287.50	0094	0063.92	0062.04	0037.60
0095	0169.60	0233.45	0288.85	0095	0064.60	0062.70	0038.00
0096	0170.72	0234.80	0290.20	0096	0065.28	0063.36	0038.40
0097	0171.84	0236.15	0291.55	0097	0065.96	0064.02	0038.80
0098	0172.96	0237.50	0292.90	0098	0066.64	0064.68	0039.20
0099	0174.08	0238.85	0294.25	0099	0067.32	0065.34	0039.60
0100	0175.20	0240.20	0295.60	0100	0068.00	0066.00	0040.00
0200	0241.20	0306.20	0363.60				
0300	0307.20	0372.20	0431.60				
0400	0373.20	0438.20	0499.60				
0500	0439.20	0504.20	0567.60				
0600	0505.20	0570.20	0635.60				
0700	0571.20	0636.20	0703.60				
0800	0637.20	0702.20	0771.60				

USE AT&T series 2000/3000 channel mileage charges can be determined from these charts. For channel segments of 100 miles or less,
part I is used. For channels of greater than 100 miles, the sum of the
hundreds mileage rate in part I and the appropriate tens-units mileage
rate from part II will yield the appropriate total rate.
Part II, column A, B, & C apply as follows:

if Total Mileage Is	MPL Schedule	I	II	III
101 - 1000 miles		B	B	A
1001 - 3000 miles		C	C	C

EXAMPLE -- 683 miles for a schedule II MPL channel segment

600 miles from part I	$570.20
83 miles from part II, (B)	54.78
Total IXC charge	$624.98

1976 Center For Communications Management, Inc   Issue Date Oct 15 1976

equipment so that they can be used as a single high-speed channel. Another way of using **TELPAK** is to derive low-speed channels out of the voice grade channels. Each voice grade channel
can be subdivided into two low-speed channels. Figure 7-20
shows the costs of utilizing **TELPAK** lines. Government agencies

IXC Mileage Rates (per mile per month)			
Type 5700 $32.50 (60 voice grade lines)			
Type 5800 $92.00 (240 voice grade lines)			

Service Terminals (per month)			
Wideband Service	Installation	First in End Office	Next in End Office
40.8 kbps/50 kbps*	$216.00	$460.00	$324.00
19.2 kbps	$216.00	$460.00	$406.00

*kbps—kilo bits per second

Figure 7-20: TELPAK (Series 5000) Costs

and certain firms whose rates and charges are regulated by the government may share **TELPAK** services. However, a private corporation cannot lease a **TELPAK** service and sell the excess lines to other companies. Such selling would put a corporation in the position of being an unregistered common carrier under the rules of the Federal Communications Commission (FCC). This service soon may be discontinued.

Wideband (Series 8000) data transmission facilities are offered as a Type 8800 line. The line is a 48 kHz carrier which is used for high-speed data transmission, facsimile transmission, or alternatively as 12 individual voice grade lines. Also, this line may be used simultaneously for both voice and data transmission up to its capacity of 12 voice grade lines. A series 8000 line can also be arranged to accommodate any two of the following three types of transmission: (1) up to six voice grade lines, (2) 19.2 kbps data transmission, or (3) facsimile transmission. Figure 7-21 lists wideband (Series 8000) costs.

Digital service is a communication facility for the transmission of synchronous data signals. The service is offered for two point and multipoint arrangements in either FDX or HDX transmission modes. Its purpose is the same as Series 2000/3000 voice grade service but it is more error free.

The digital transmission network employs bipolar pulse code modulation techniques. It contains extensive diagnostic features for error detection and correction. The local loops require special treatment to enable a data range of 56 kilobits per second between the end office and the local customer's premises. Regenerative repeaters are employed every several thousand feet in the

IXC Mileage Rates (per mile per month)			
Wideband Service	First 250 Miles	Next 250 Miles	Per Mile Over 500
Type 8800	$16.20	$11.40	$8.15

Service Terminals (per month)			
Wideband Service	Installation	First in End Office	Next in End Office
40.8 kbps/50 kbps*	$216.00	$460.00	$324.00
19.2 kbps	$216.00	$460.00	$406.00

To provide for use of the type 8800 line as 12 individual voice grade lines, add $271. per month to the service terminal charges above.

*kbps—kilo bits per second

Figure 7-21:   Wideband (Series 8000) Costs

local loops to refresh the digital pulse train. This approach also eliminates the need for complex analog modems. The average bit error rate in the digital network is as low as one error for every 10^7 bits transmitted. American Telephone and Telegraph claims that 99.5% of 1 second time intervals are, on the average, error-free. Furthermore the digital network has a channel availability of at least 99.96%.

The digital network is still growing but it will be available only in 100 to 200 of the large metropolitan cities of the United States. Digital service may not be available in areas of the United States that are served by small independent telephone companies. In other words, a user of digital service between two locations not in the digital service network would have to obtain regular voice grade line service at each end of the digital network. Examples of the currently served cities are Atlanta, Baltimore, Boston, Chicago, Cleveland, Dallas, Denver, Detroit, Hartford, Houston, Kansas City, Los Angeles, Miami, Milwaukee, Minneapolis, Newark, New Haven, New York, Philadelphia, Pittsburgh, Portland, San Francisco, Saint Louis, and Washington, D.C. The total monthly cost of digital service is composed of three items: the intercity line mileage, the service terminals, and the network interface equipment. There is no rate difference between FDX and HDX and the digital service is on a 24-hour per day basis. Figure 7-22 shows the costs of digital service.

Item*	Speed			
	2400 bps	4800 bps	9600 bps	56000 bps
Mileage charges	$21.00 plus $0.41/mile	$41.00 plus $0.62/mile	$62.00 plus $0.93/mile	$129.00 plus $4.12/mile
Service Terminals Channel Termination (per channel end)	$10.00	$21.00	$31.00	$65.00
Data Access Line (local loop) Type 1: <5 miles from end office	67.00	88.00	113.00	206.00
Type 2: >5 miles from end office	93.00+ 0.62/per mile	113.00+ 0.93/per mile	134.00+ 1.34/per mile	258.00+ 6.18/per mile
Network Interface Equipment Data Service Unit† Channel Service Unit‡	16.00 0	16.00 0	16.00 0	21.00 0
One Time Installation Costs Data Access Line Data Service Unit Channel Service Unit	$103.00 26.00 0.00	$103.00 26.00 0.00	$103.00 26.00 0.00	$155.00 26.00 0.00

*These are per month charges except for the installation costs.
†Data Service Unit affords a RS232C interface plug
‡Channel Service Unit terminates the line (no modem required) and is used when the user equipment provides timing recovery, bipolar code conversion, and other control logic.

Figure 7-22:   Digital Service Costs

Satellite service is a point-to-point communication facility for voice, data, facsimile, and various other wideband applications. It is furnished through domestic satellites in combination with conventional land line access facilities. The service is offered on a 24-hour per day basis, as are leased voice grade lines. All satellite lines are four-wire circuits so they can be utilized for FDX transmission. Voice grade satellite channels are available only to major cities. The use of extended land line access facilities, e.g., voice grade—Series 2000/3000, is necessary to reach smaller cities. Examples of major cities are Atlanta, Chicago, Dallas, Houston, Los Angeles, New York, Pittsburgh, San Francisco, and Washington, D.C. The cost of utilizing satellite channels is shown in Figure 7-23. The total monthly cost for a satellite line is

Satellite Channel Charge (per month)			
Voice Grade Channel			
		Number of Channels	
Mileage	1–5	6–11	12–23
500–1089	$ 500	$450	$400
1090–1889	750	675	600
1890 & over	1000	900	800
Wideband Channel			
Mileage	48 kHz	240 kHz	1.2 MHz
500–1089	$5,100	$21,000	$ 90,000
1090–1889	6,750	24,750	123,750
1890 & over	9,000	33,000	165,000

Channel Terminal per Voice Grade Channel (per satellite access only)

Installation	Monthly
none	$20.00 each

Service Terminal per Voice Grade Channel (per station termination)

Installation	Monthly
$50.00	$25.00

Figure 7-23:   Satellite Channel Costs

composed of three items: satellite channel charge, channel terminal, and service terminal. In addition, satellite lines can be
conditioned to reduce the number of errors during transmission.

Cost of Other Communication Equipment

This section offers the user examples of various equipment costs.
These are representative costs that can be used when developing
network designs or when answering exercises and network problems posed in this textbook. These costs are representative only
and are subject to variation with time. When doing detailed
analyses, consult vendors for exact prices. Figures 7-24 to 7-26
list the costs of various equipment.

Questions—Chapter 7

A critical stage in the design of any data communication
system is the process of cost-optimization. Customers have to be

Transmission Speed	Installation Cost	Monthly Cost
0–300 bps †	$ 30.00	$ 25.00
2000 bps	85.00	60.00
2400 bps	85.00	60.00
4800 bps	165.00	135.00
9600 bps ‡	220.00	250.00

†bits per second
‡May require line conditioning

Figure 7-24: Examples of Modem Costs

Type	Installation Cost	Monthly Cost
Teletype (0–75 bps)	$55.00	$150.00
TWX (60 wpm)	50.00	125.00
TWX (100 wpm)	50.00	160.00
Video (2400–4800 bps)	50.00	150.00
Printer for video terminal	25.00	70.00
Cluster controller for 6 video terminals	150.00	200.00
Intelligent terminal	100.00	200.00–800.00

Figure 7-25: Examples of Terminal Costs

Teletypewriter Line Concentrators:	Installation Cost	Monthly Cost
22 stations to 5 channels	$110.00	$ 60.00
200 stations to 30 channels	270.00	60.00
FDM Multiplexers 75 and 150 bps channels from series 3000 circuit. Up to 17, 75 bps or 8, 150 bps channels or combinations thereof		
Multiplexer Up to 3 channels	220.00	200.00
Each additional channel	110.00	35.00

Figure 7-26: Examples of Multiplexer (MUX)/ Concentrator Costs

served at a reasonable cost, even if this sometimes means that hoped-for service levels have to be compromised somewhat. Thus, the designer must competently assess the costs of proposed system alternatives. To do this, he must be aware of all service offerings and their relative costs. The questions that follow are designed to help you measure your ability to meet these criteria.

True or False

1.   Low-speed private leased lines are used for transmission speeds up to approximately 9,600 bits per second.

2.   WATS service for a band from California that would include New York would cost approximately $55 per month.

3.   The cost of a modem generally exceeds $500 per month.

4.   There are six types of leased line communication services.

5.   All communications common carriers in the United States come under the jurisdiction of the FCC.

6.   Some communications common carriers offer a broad range of services while others offer only specialized services.

7.   On any link of a communication network, the transmission speed is usually solely dependent on the type of line chosen.

8.   Band 5 or A-3 WATS service would give a California subscriber a calling area including the entire continental United States.

9.   TWX allows transmissions at either 60 words per minute or 100 words per minute.

10.   50 Kilobit service is classified as a leased communication service.

Fill-in

1.   When WATS is offered on a measured service basis, it includes ———— hours per month in the basic charge.

2.   When WATS is offered on a full business day service basis, it includes ———— hours per month in the basic charge.

3.   ——— ——— is a metered use, private line service between two points only.

4.   Voice grade leased line charges are on a per ——— per ——— basis.

5.   IXC stands for ——— ———.

6.   Satellite service is offered on a ——— hour per day basis.

7.   In packet switching, data are generally segmented into ——— character blocks.

8.   There are ——— types of switched communication services.

9.   Wideband data transmission can accommodate several voice grade lines, 19.2 kilobit/second data transmission, or ——— transmission.

10.   CCSA stands for ——— ——— ——— ———.

Multiple Choice

1.   Which of the following is an example of a switched communication service?

a) CCSA
b) Digital service
c) Direct distance dialing
d) A and B above
e) All the above

2.   Which of the following would be considered a hybrid communication service?

a) TELEX
b) TWX
c) Foreign exchange
d) All the above
e) None of the above

3.   Which of the following does *not* base its charges at least partially by distance?

a) DDD
b) 50 Kilobit
c) TWX

    d) Packet switching
    e) All the above have charges based on mileage

4.   TELEPAK lines provide maximum channel capacity equivalent to:

    a) 2 voice grade lines
    b) 60 voice grade lines
    c) 60 or 240 voice grade lines
    d) 5,000 voice grade lines
    e) None of the above

5.   Voice grade channels have a bandwidth of:

    a) 150 cycles per second
    b) 960 cycles per second
    c) 3,000 cycles per second
    d) 9,600 cycles per second
    e) 240 kilohertz

6.   Wideband lines provide a maximum channel capacity equivalent to:

    a) 2 voice grade lines
    b) 60 voice grade lines
    c) 60 or 240 voice grade lines
    d) 5,000 voice grade lines
    e) None of the above

7.   Digital service employs the use of all the following *except:*

    a) Bipolar pulse code modulation techniques
    b) Complex analog modems
    c) Regenerative repeaters
    d) Synchronous data signals
    e) All the above are normally used with digital service.

8.   Which of the following is considered a leased communication service?

    a) 50 Kilobit
    b) Hot line
    c) Metered WATS
    d) Satellite
    e) None of the above

9.   Which of the following type of service is restricted to the fewest cities at the present time?

    a) 50 Kilobit
    b) Packet switching

c) Digital service
d) WATS
e) Satellite

Short Answer

1.   Carefully draw a graph to scale depicting Band 5 Northern California WATS rates for usage of 0 to 300 hours per month. Plot both the measured and full business day rate schedules on the same set of axes. Estimate the hours of usage that correspond to the breakeven point between the two services.

2.   Calculate the breakeven point in Problem 1 by solving the following equation for H (the breakeven usage in hours).

$$\text{Measured time cost} = \text{full business day cost}$$
$$255 + 19.10 \, (H\text{-}10) = 1,675$$

Why is the hourly rate multiplied by H-10?

3.   For a 100-word-per-minute TWX service, using characters consisting of 7 bits plus parity, 1 start bit, and 2 stop bits, if a word is 6 characters, what is the transmission rate in bits per second?

4.   What is the difference between interstate and intrastate data communications?

5.   What are the switched services that were discussed in this chapter?

6.   What would be an ideal block length when using synchronous transmission over a packet switching network?

7.   Which would cost more—a Series 1006 line or a Series 2000/3000 line between San Francisco and Los Angeles? (Assume the distance to be 347 miles).

8.   Assume that you have two very small towns that are 100 miles apart. Which would cost more, a Series 1006 line or a Series 2000/3000 line?

8

Designing Communication Networks

The preceding chapters of this book have laid a factual
and methodological foundation. This chapter is the
payoff. The fundamental problem in designing
data communication networks is not whether the
technology is available but how best to adapt its
offerings to meet the changing and challenging
needs of business and government
organizations. The conversion of ideas to
reality requires the use of methods and resources. Thus,
the systems designer must define the practical
methods and the proper resources that will be required
to move the system from the conceptual stage
to actual operation.

An Overview

Many businesses today operate in numerous locations. It is not
unusual for an organization to have tens, hundreds, or even a
thousand or more sites at which transactions take place. Management of such organizations must have current knowledge of
their geographically scattered operations in order to best serve
their customers, meet the competition, and maintain close surveillance over critical activities. This calls for the rapid collection, processing, and distribution of business information.

Advances in computer design, remarkable reductions in cost per computer operation, and creative ideas in computer applications have brought about increased use of data communication systems for conveying information between widely separated business locations and the computers and terminal equipment installed at these locations. Thus, it is possible for the management to know in seconds what the state of affairs is at a branch or at any other location throughout the country or the world.

This chapter is the culmination of all the information in this book. Up to this point, you have learned about fundamental data communication concepts, equipment, software, and tariffs. To design a data communication network, the designer must put this information together creatively and systematically. The intent of this chapter is to provide guidance in being systematic, and stimulation in being creative. During the design process you should refer back to earlier sections of this book to refresh your memory, clarify your understanding, and, in general, do a thorough job.

The Systems Approach to Design

The systems approach to a design problem differs from the trial-and-error approach. True to its name, in the systems approach we try to identify all influences and constraints on the desired result of the design and evaluate them in terms of their impact on the processes, hardware, software, and people that make up the system. There are many ways to do this, but all can be defined in terms of 10 basic design steps. First, we give a brief description of the 10 basic steps, and then discuss each step in detail.

Step 1: Define the problem so it is evident and understandable to management, the potential users, and the system design personnel. All these parties must work toward a solution to the same problem. The solution should be stated in a positive manner as a course of action that management can take.

Step 2: Prepare an outline of the approach and methodology that you are going to utilize in studying and designing the proposed system, in order to organize a detailed plan of action. The outline should be based on Steps 3 through 8 of this sequence, adjusted to the particular problem in hand.

Step 3: Gather general background information on the areas that will be affected by the proposed data communication system. It is imperative that the system designer know the background of the industry, company, or governmental agency, and the various individual areas that will be affected because every aspect of the business is integrated to work together to accomplish the organizational objectives.

Step 4: Study the interactions between the areas that are affected by the proposed data communication system. Learn the interactions between the various departments or agencies within the organization. These interactions should be defined in terms of the outputs and inputs of each organizational unit and the processes performed by the various organizational units, insofar as they affect, or are affected by the design.

Step 5: Obtain a general understanding of the existing system. The designer must understand the existing system, whether it is a manual system or an on-line data communication system that is being replaced or enhanced. It is important in the design process for the designer to know why certain things are done in the way they are.

Step 6: Define the proposed system's requirements in order to assemble an overall picture of the system. These requirements must be defined within the framework of the goals and objectives of the entire organization as well as of each department or agency. Strive to make these requirements quantitative and detailed.

Step 7: Using the requirements defined in Step 6, design the proposed data communication system. If the breadth of the requirements allow it, design various alternative systems, taking different technical and/or procedural approaches. Consider the maintenance and network management aspects of each alternative system.

Step 8: Develop the cost comparisons for the various alternative designs. Select the network configuration alternative that best meets both the cost and system requirements. Make minor adjustments to optimize the design, if possible.

Step 9: Sell the system. In this important step, the system designer must convince both management and the users that it is cost-effective to implement the data communication system design and that it will produce the desired results.

Step 10:   Implementation, follow-up, and reevaluation are mandatory steps if the system is to be widely accepted by its users. In the implementation process, the system designer participates in the procurement, development and installation of the system, as well as in organizing the maintenance procedures and training its users. During follow-up, the system designer observes the actual operation and makes sure that all parts of the new system are operating according to specifications. Reevaluation takes place after the system has been broken in; the system designer returns to evaluate its efficiency and make whatever changes are needed to optimize its performance.

Notice that these 10 steps could apply to most any system, not just to data communications. The remainder of this chapter is a discussion of these steps as they apply directly to the development of data communication systems.

Identify and Define the Problem (step 1)

A successful systems designer will always start by defining the problem at hand prior to working toward its solution. Devote an adequate amount of time toward problem definition because often what appears to be the problem itself may be only a symptom of the real problem.

Suppose, for example, the sales manager calls in the communications manager and complains about a slow order-entry system that results in many lost sales. At first glance, the solution might seem to be a high-speed data communication network, but further probing may show that the real culprit is a complicated and cumbersome order entry form. In other words, a high-speed data communication network would not do much to solve this problem unless the format of the order input form were redesigned simultaneously.

The objectives to be achieved by the proposed data communication system should be described. The objectives should be stated in terms of the organization's own objectives. The system designer should clearly define and analyze the ramifications of the problem, point out interrelationships between the problem at hand and any other problems, activities, or situations uncovered while identifying the problem, state the objectives to be met as well as the scope of the project, and indicate the areas of the organization that are included in or excluded from the problem definition. The result of this work should be documented in a formal report, which should explain the logic used to arrive at the problem definition and the analyses. The report should also include any charts, graphs, pictures, floor plans, layouts, or map configurations that are required to clearly define the problem.

Prepare an Analysis and Design Plan (step 2)

Once the problem is defined, the system designer can prepare an outline of the plan of attack. This plan should be the roadmap that the system designer utilizes to conduct the rest of the design efforts. This plan should include the activities necessary to perform Steps 3 through 8 of the systems design approach. The plan should identify:

- Sources of information to be used
- Types of information to be collected
- Analyses to be performed
- Schedule of all activities
- Definition of results to be produced
- Schedule of time and cost

This plan should be approved by your supervisor since assistance will probably be needed to smooth the way for some of the data gathering. Subsequently, it should be shown to other concerned individuals such as the system's users. Do not underemphasize this step—poor planning is a primary reason for many system failures.

Gather Background Information on Affected Areas (step 3)

While data communication systems are products of human technological achievement, they are still operated by people. Data communication systems have many similarities, but they have unique characteristics as well. These unique characteristics are important to the system designer. The system designer must gather and analyze the facts to determine the specifications for the distribution of information throughout the data communication network. Therefore, the designer should gather background information on the organization as well as on the various departments or agencies that will be affected by the data communication system. If you are unfamiliar with the industry or governmental structure, study it also. Find out what competitors are doing or what similar government agencies are doing. Determine if there are any legal requirements or organizational policies that will affect the future design of the network. A technically feasible (even optimal) design may not be organizationally or legally acceptable.

For example, in the case of the proposed order entry system, we must be sure that management's policies covering priorities, commissions, payment terms, and other "customer relations" matters are properly dealt with, even if this means communicating much more data (thereby incurring more costs) than is necessary for ordinary order entry functions.

Study Interactions Between Affected Areas (step 4)

In all likelihood, the proposed system will have important effects outside the mainstream application area. For example, the sales manager's new order entry communication system may shorten lead time to delivery, provide more immediate knowledge of what is in the order pipeline, and generally change some of the parameter values that manufacturing has depended upon in recent years. Make an exhaustive list of potential interactions between affected areas, departments, or agencies. Evaluate each possibility until you know the extent and impact of the interaction or until you determine that it is insignificant. Factor the significant interactions into the remainder of the design effort.

Understand the Existing System (step 5)

Understand the system that is currently in place. This will provide a benchmark for measurement and a point from which to start your design efforts.

Determine the current pattern for internal distribution of organizational information. Draw a preliminary chart showing all the locations that receive or transmit information. These charts may take many forms. One form is a set of charts, one for each department or agency; another, is an overall chart that shows the interconnection of the various departments or agencies within the entire organizational structure. The locations are the various entities of the organizations that provide the essential operating information. These locations should be identified as to the type of information that is exchanged (e.g., inquiry-only or input/output) plus a general description of the types of information transmitted from each location. Finally, diagram the information flow on a chart. If the communication system is geographically dispersed, a map may be required.

Define the System Requirements (step 6)

Gathering and analyzing the facts to determine the distribution of information is primarily a systems analysis function. A flowchart is extremely helpful when looking at any information system. The data communication designer should note that if the present method of performing the functions can be significantly improved, then perhaps automation is unnecessary. The point to be made is that if a system is inefficient or does not satisfy the current user's needs, then a mirror-image automated system may not satisfy the user's needs either. If the problem has been defined adequately, this step will be easier.

The system designer now has a problem definition, the necessary background information, knowledge of interactions between affected areas, and a general understanding of what the communication system is to accomplish, as well as a basic understanding of the business systems that are to utilize the data communication system.

The system designer is now ready to begin the exacting and detailed work that is unique to the design of a data communication network. This work begins with the definition of system requirements. This is the most critical step of the whole process. The seeds of failure are more often planted here than at any other point. The most important caveats are:

- Distinguish between requirements and desirable features. Be sure you know and state both what is mandatory and what would be a desirable item. It is important to list desirable items, but label them as such.

- Be quantitative, precise, and performance-oriented. "A minimum of 1,000 messages of 150 characters average length in the peak hour" is far better than "a large number of messages."

- Distinguish between *requirements* and *solutions*—solutions do not belong here. "Printing should be easily readable by a person with normal eyesight under normal office lighting conditions" is a requirement. "Printing shall be 0.1 inches high in italic font" is not a requirement; it is simply a solution masquerading as a requirement. The designer who does not make this distinction must accept all responsibility that the solution will solve the underlying problem, although in many cases it probably will not.

- Be sure to include requirements for:
  —Inputs/outputs
  —Processing

—Hardware/software
—File structures/data base
—Common carriers
—System monitoring/management
—Interfacing with other systems
—Documentation
—Reliability
—Accuracy
—Maintainability
—Safety
—Security/privacy
—Human factors
—Training
—Pre-cutover testing
—Implementation
—Future growth

When you have documented the requirements for the proposed system, secure the approval of all users, management, and the EDP/communications operations. Once all parties are agreed that the requirements are correct and complete, the actual design may be started with confidence that the right problem is being solved.

Design the Proposed Data Communication System (step 7)

This step can be broken down into seven substeps:

- Analyze the types of messages/transactions
- Determine message lengths
- Determine message volumes
- Determine total traffic
- Establish line loading
- Develop alternative configurations
- Consider hardware and software requirements.

Because these substeps are at the heart of the design process, we will discuss each one individually.

Analyze the types of messages/transactions The system designer must now determine the exact nature of the messages to be transmitted. This includes both the generic type of message and the information elements of each message to be transmitted. Some messages may be short inquiries requiring a short response,

Message Number	Message Name	Estimated Number of Characters
1	Weekly payroll timesheets	45
2	Daily inventory update	90
3	—	—
4	—	—
5	—	—
	etc.	

Figure 8-1:   Message List

or inquiries evoking a longer reply. Others may be rather lengthy input messages that are involved with the day-to-day business operations of the organization. These may be responded to with a brief acknowledgment or with a lengthy answer.

Identify each message by a short title and gather a sample of the message if there is now an equivalent. If there is no current equivalent, then the system designer must design the message and have it approved by the appropriate user departments or agencies that will be utilizing the data communication network for their day-to-day business activities.

In either case, first list all the items of information to be transmitted and estimate the number of characters to be allotted for each message. Make a list, such as the one illustrated in Figure 8-1, showing each message type. The Message List should be verified by the users of the basic business system involved. At this point, the system designer and other system analysts from the organization may have to reexamine the underlying business system because of questions raised by the message analysis. These questions should be settled if at all possible before proceeding. You may be forced to backtrack in the design process if substantial agreement cannot be reached at this time.

Determine message lengths: Calculate, as accurately as possible, the average number of characters for each element of information in each message type you have included in the Message List (Figure 8-1). Develop descriptions of specially designed messages and/or collect samples of the messages that are in the current system. Expand the data of the Message List into the form shown in Figure 8-2. Upon completing this step, the designer should have a table that lists the average number of characters and the maximum number of characters by message type.

The system designer should note that a pure character count may be misleading for estimating the number of characters being transmitted. For example, if a message of 120 characters is sent to a video terminal, the actual characters on the data line could approach double that when "null," "blank," "sync," carriage return, line control, etc. characters are included.

Determine message volumes: The system designer must calculate various statistics about the daily or hourly volumes of messages flowing in the system. In addition to the average daily volume, the designer must determine the peak daily (hourly) volumes, and any seasonal variations. (Look ahead to Figure 8-4).

There may be factors causing peak loads in the data communication system on a daily basis. Determine whether there are peak loads that will affect the system during various times of the business day, week, or month. Another question to be asked is whether there is an increase in the number of characters per message during peak volumes of the business day. Determine if there will be any seasonal characteristics that might affect peak loads of the data communication system.

The designer should plan for varying volumes at different hours of the day. For example, in a banking system, the network

Message Number	Message Contents	Average Characters/ Message	Maximum or Peak Characters/ Message
1	Name	20	25
	Employee number	6	6
	Regular hours	5	5
	Overtime hours	4	5
	Shift	1	1
	Special status	0	20
2	Part number	9	9
	—		
	—		
	—		
	etc.		

Figure 8-2: Message Types

designer could plan on the bank opening at 9 A.M. with some limited traffic between 9 A.M. and 10 A.M. as tellers prepare for work. From 10 A.M. to about 12:30 there will be a peak of message activity. Between 12:30 and approximately 2 P.M. there will be an average volume, and from 2 P.M. until 3 P.M. there will be a second peak of business activity. After 3 P.M. (the bank's normal closing hour) there will be an average volume of transmission consisting primarily of internal bank business between branches. If the bank is open on Friday evening or Saturday, this, too, will require consideration.

In another case, an airlines reservation system might expect some rather severe peaks in message activity just before and during the Christmas travel season. Peaks such as this might be so irregular (happening only once a year) that the system designer cannot design the system to accommodate this peak activity without raising the cost to an unreasonable level.

The system designer can use the present system (it may be a manual system or the current data communication network) to calculate the average daily volumes. The system designer should take a random sample of several days' traffic and actually count the number of messages handled each day at each location. The designer should ensure that the average message count, the peak message count, and an example of all the different types of messages that are currently flowing in the system are taken into account. The working days selected for sampling should be chosen at random, unless there are systematic variations on the days of the week. If possible, the designer should also take into account any seasonal variations. For organizations that have daily variations, the designer should take an average message count for both a normal period and a peak period. The count should be taken at each location from which information is sent and each location at which information is received.

If the system from which the average daily volumes are required is nonexistent (even in a manual form), then the system designer must use estimating procedures to determine the average daily volumes and peak volumes of messages that will be flowing in the system. There are three kinds of estimating procedures that are generally useful in systems design work. These are conglomerate estimating, comparison estimating, and detailed estimating. In conglomerate estimating, representatives from each functional area that will be utilizing the data communication system meet to develop average and peak message volume estimates based on past experience. In comparison estimating, the system designer meets individually with anyone, inside or outside the organization, who has a similar system. The designer then evaluates comparable operations and develops average and peak volume estimates. In detailed estimating, the designer makes a detailed study of the messages, their movement, the information they must contain, and any other pertinent factors for each step of each procedure within the system, and then calculates the average and peak message volumes. When extreme accuracy is required, perhaps because of the importance of the decisions involved, detailed estimating should be used.

Determine total traffic: Now that the designer has the average/ peak message volume and the average/peak characters per message, the total traffic in characters or bits per second should be determined on the basis of total characters transmitted per day or hour.

The system designer should go back to the preliminary geographical map or locational distribution chart that was developed earlier in the system study (Step 5). This map or chart should be reviewed to determine if it still seems reasonable in light of the further information that has been gathered and calculated to this point. If not, the system should be reconfigured as appropriate. In other words, the designer should now have a rough draft network configuration map that shows connecting lines between the central computer and each of the various input/output stations that will be connected to the computer. This type of a network configuration map is shown in Figure 8-3.

The network designer is now ready to ascertain the total traffic that will be transmitted over each link of the network. These data are consolidated into a table, as shown in Figure 8-4. List each of the network links in Column 1 such as Miami to Tulsa or Los Angeles to Tulsa. For each link, list in Column 2 all the message types that link will transmit, (e.g., weekly payroll, order entry, accounts receivable). Double Columns 3 and 4 contain the average and peak number of characters per message and the average and peak messages per day for that particular network

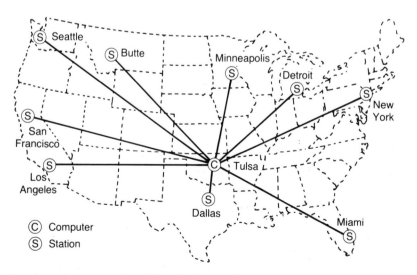

Figure 8-3: Geographical Distribution of Stations

(1) Network Link	(2) Message Type	(3) Characters/Message		(4) Messages/Day		(5) Characters/Day		(6) Hourly Number of Characters Transmitted										
		Average	Peak	Average	Peak	Average	Peak	8-9	9-10	10-11	11-12	12-1	1-2	2-3	3-4	4-5		
Miami/ Tulsa	1	36	62	20	20	720	1240							720 to 1240				
Totals																		

Figure 8-4: Network Link Total Traffic

link. Multiply Column 3 by Column 4 to obtain the total average and peak characters transmitted per day, and insert these figures in Double Column 5.

If the hour-to-hour variation is significant, the total characters transmitted per day must be developed into an hourly number of characters transmitted. To do this, the system designer reviews each transmission link and performs an hour-by-hour traffic study of the number of messages transmitted. The right half of Figure 8-4 can be utilized for recording this type of information.

After the total characters transmitted per day per link are calculated, they should be converted into bits per second to determine what type of transmission line should be utilized to connect each station to the central computer. For example, if 200,000 characters per day are transmitted between Miami and Tulsa utilizing an 11-bit per character code (asychronous transmission), then the line speed will have to be 77 bits per second assuming an 8-hour work day. Figure 8-5 shows the calculation of the 77 bps. Note that in this example, no allowances were made for contingency factors such as inefficiencies, outages, uneven demand, polling, line turnaround, retries due to errors, and the like. Multiplexing and other types of multipoint line control may be considered at this point in the design process (Chapter 3 discussed multiplexing and Chapter 4 multipoint line control).

The total characters transmitted per day or per hour must be calculated for each network link by utilizing a form similar to that shown in Figure 8-4. Some designers record the total characters or bits transmitted per day on their geographical map (Figure 8-3) for each link in the network.

Establish line loading: To establish the line loading (the amount of data transmitted) the designer must start with either the total characters transmitted per day on each link, or the number of characters transmitted per hour during the business day on each link of the network. Starting with the total characters transmitted per day, the system designer first determines if there are any time zone differences between the various stations. In the previous example, there is a two-hour time difference between the San Francisco station and the central computer in Tulsa (see Figure 8-3). This means that if the computer at Tulsa is activated at 8 A.M. Tulsa time, it is only 6 A.M. in San Francisco; if Tulsa shuts down its computer at 5 P.M. it is only 3 P.M. in San Francisco. If the Tulsa computer site shuts down at 3 P.M. San Francisco time, the effective work day for San Francisco is shortened to six hours (assuming a one-hour lunch). This may cause a difference in the

```
200,000  characters/day
×    11  bits/character code
─────────
2,200,000  bits/day
÷     8  hours
─────────
  275,000  bits/hour
÷    60  minutes/hour
─────────
    4,584  bits/minute
÷    60  seconds/minute
─────────
       77  bits/second
```

Figure 8-5:   Conversion of Characters Per Day to Bits
Per Second

speed at which data must be transmitted or it may cause nonstandard work schedules to be followed at one or both ends of the link. For example, if the work day was shortened to six hours in Figure 8-5, the line would have to transmit at 102 bps instead of 77 bps. An alternative to this would be to have a terminal operator start work at 6 A.M. in San Francisco, if this was compatible with the type of business, or schedule staff during the lunch hour in San Francisco, to gain one more hour.

Other factors must be considered when determining line loading. One major item is the efficiency of the code utilized for transmission and the efficiency of the method of transmission (asynchronous versus synchronous). Examine the effectiveness of various codes, e.g., whether a 9-bit-per-character code will be as effective as an 11-bit-per-character code.

The computer growth factor must be considered at this point because the useful lifetime of a data communication system should be at least two, and preferably as many as five, years to justify its intitial cost. Forecasts should be made of the message volumes that will be faced two to five years in the future. This growth factor usually varies from a 10% to 50% increase in the number of messages transmitted.

The designer should also allow some time for transmission line errors (error detection and correction) resulting in the re-transmission of the same message. The system designer should probably consider a 2% to 10% contingency factor for retransmission of messages received in error.

The system designer must take the "turnpike effect" into consideration. The turnpike effect results when the users utilize the system to a greater extent than was anticipated because the system is found to be efficient, available, and quick.

Another area of concern is the use of a priority message scheme. The system designer may have to plan for a greater throughput to ensure that low-priority messages can get through the network without inordinate delays. Sometimes high-priority messages so overwhelm the low-priority messages that the low-priority messages are not transmitted within a reasonable length of time.

Some other contingency factors that must be taken into consideration are the terminal operators' learning curve when learning the new system, inaccuracy of the traffic analysis that the system designer performed, and the effect of business operating procedures on the system itself. Besides the traffic analysis, there may be many other forms of estimation errors for which the designer should allow a small contingency factor and increase the minimum speed at which the data are transmitted.

Other items that utilize transmission time, but do not transmit business data, and for which the designer must account are:

- Polling time
- Line control characters
- Turnaround time
- Modem synchronization time
- Message propagation time
- Printer time (form feed, tab, carriage return)
- Keyboard time (in conversational mode)
- Card or tape reading time (if unbuffered)
- Code conversion time (if software mediated).

At this point, the system designer should review and establish the response time criteria that are required to meet the requirements of the data communication system being designed.

The problem takes on a different dimension if, for a while, the demand for transmission capacity exceeds the capacity available. This can happen in several ways. One such occurrence is when line noise causes repeated retransmission of a few messages and other traffic "backs up" (the same situation as when an accident happens during rush hour on a freeway). A more common cause is a short-term peak in the demand to transmit by a number of terminals that share the same facility (multipoint line or front-end communication processor receiving port). In this case, the analogy is to a pair of freeways that merge into one. If, for a short period, more cars arrive at the merger of the two incoming roads than can be carried off by the outgoing one, some cars must pause, and traffic will back up on the incoming roads.

There are ways to determine the likelihood that a traffic buildup will occur, and to predict how serious it will be. These methods use "queuing theory," a fairly abstruse mathematical concept that is beyond the scope of this book. The system designer, however, should be alert to the possibilities of this type of line loading. Ask if the rate of production of inputs could ever exceed the line capacity. If so:

- What is the statistical distribution of the message lengths?
- What is the statistical distribution of the intervals between messages?
- What is the longest transmitting delay that can be tolerated?
- What is the maximum number of backed-up messages that can be tolerated?

With the answers to these questions, the system designer can refer to one of several excellent texts* to determine whether extra capacity must be designed into the system to reduce the probability that backups will occur, or whether some compromise in system performance must be accepted.

Develop alternative configurations: At this point, the system designer utilizes all the information collected to date (especially the network configuration map or charts) and seeks ways to configure the network at minimum cost. Review the map that links the various station locations and the computer. If the nature of the problem permits, move these stations about and, in any case, try various line control methods to arrive at the lowest cost. To achieve this, two items can be manipulated: IXC mileage, and line control method combined with channel type. Assuming that the various stations are in fixed locations, the designer must evaluate various types of line control methods. For example, connecting each of the stations to the central computer on a point-to-point basis is the simplest, though probably the most expensive approach, because of excess capacity in some lines. Figure 8-6 shows such a point-to-point configuration represented by the dashed lines. An alternative is also shown, using a multiplexer in Detroit to combine the traffic from Boston, New York, and Detroit into one broader band line to Tulsa (represented by solid lines). Other cities are handled similarly, but note that no good solution exists for Miami and Minneapolis. Finally a possible configuration using multidrop lines is indicated in Figure

---

*For example, James Martin, *Systems Analysis for Data Transmission* (Englewood Cliffs, N.J.: Prentice-Hall, Inc., 1972), chapter 31.

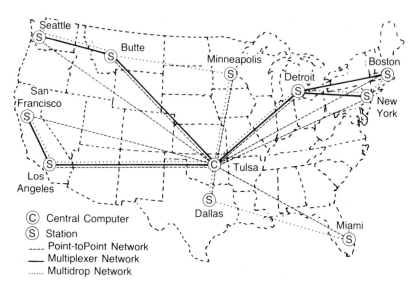

Figure 8-6:   Comparison of Line Control Methods

8-6, by the dotted lines. There are similarities to the multiplexed network, but note that Minneapolis is now served more economically. Both the multiplexer and multidrop configuration assume that the point-to-point lines had enough excess capacity to handle the messages of the other stations added to these lines.

The designer should consider dial-up stations for low volume terminals, especially if it can be evening dial-up, which is inexpensive. Multiplexer/concentrator configurations are usually adaptable to areas generating high traffic. In any case, the objective is to configure the system so there is a minimum cost for IXC mileage plus terminal equipment, and so the lines can carry the number of bits per second that are required to conduct the daily business.

As an example of minimizing both line mileage costs and equipment costs, suppose the data communication system designer at Fitson, Inc., is confronted with the following problem:

> Fitson, Inc., is going to open a new regional sales office in New Orleans. Initially, six low-speed (10 characters per second) terminals are to be installed there, which will tie in to the order entry system that operates on the Fitson computer at corporate headquarters in Atlanta. Because high frequency of use is expected at these terminals, both WATS and dial-up service would be excessively costly. How should the leased line communications be configured for the six terminals and for additional terminals, as the office grows?

The designer should set up two formulas showing the differential costs of separate lines and multiplexed lines:

If n = number of terminals

M_L = monthly cost of a low-speed modem pair

M_H = monthly cost of a high-speed modem pair

L_L = monthly cost of a low-speed line

L_H = monthly cost of a high-speed line

MX_B = monthly cost of a basic multiplexer pair

MX_L = monthly cost of one multiplexer line adapter pair

m = maximum number of low-speed terminals that can be accommodated by the multiplexer

C_{nS} = cost of servicing n terminals on separate lines

C_{nM} = cost of servicing n terminals by multiplexing

Then:

$$C_{nS} = n(M_L + L_L)$$
$$C_{nM} = MX_B + M_H + L_H + nMX_L; \text{ for } 1 \leqslant n \leqslant m^*$$

In this case, use the following approximate values for the variables:

$M_L = \$60$, $M_H = \$180$

$L_L = \$300$, $L_H = \$380$ (for distances this short and where terminals are in major cities, low-speed lines do not show a cost advantage)

$MX_B = \$200$, $MX_L = \$20$

m = 16

Solving for C_{nS} and C_{nM} for several values of n

n	C_{nS}	C_{nM}
1	$ 360	$ 780
2	720	800
3	1080	820
4	1440	840
8	2880	920
12	4320	1000
16	5760	1080

---

*When n exceeds m, another MX_B, M_H and L_H must be added to the network, e.g., $C_{nM} = 2MX_B + 2M_H + 2L_H = nMX_L$ for $m + 1 \leqslant n \leqslant 2m$ and so on.

This calculation shows an advantage for multiplexing for three or more terminals. If the home office was in Birmingham (140 miles from Atlanta instead of 420) the advantage would start at four terminals.

Some important observations should be made about the above example:

- If all terminals are served by a single line, as in the multiplex case, there will be some degradation in overall reliability. If the line or multiplexer fails, all terminals will fail. In the unmultiplexed case, a line failure will cause only one terminal to fail. Some "cost of reliability" factor must be included in judging whether or not to multiplex.

- Modems and multiplexers are needed in pairs; therefore, costs for pairs are included in the formulas.

- The formulas deal with monthly costs. If equipment is purchased rather than leased, these formulas require that costs of equipment, maintenance, and interest on capital be spread over the useful life of the equipment to arrive at a monthly cost.

- The multiplexer formula implies a unit that consists of a basic box capable of handling one to m line adapters that can be purchased separately as needed. This is not always the case. In some instances, line adapters come in groups of three, four, or more. The designer must modify his analysis to respond to these variations.

For all practical purposes, the technical differences between FDM and TDM need not concern a designer when faced with most problems. The technical differences do have consequences for the cost of the multiplexers, however; especially for the incremental cost to add one more line. Typically, the base cost of TDM is higher and the incremental line cost is lower. Thus, one may tend to favor TDM when growth is expected after initial installation. However, no general rule is thoroughly trustworthy; the designer is cautioned to make the trade-off cost analysis of TDM versus FDM, based on the facts of the specific situation.

Consider hardware and software requirements: While the designer is utilizing various line control methods to minimize line costs, he should also evaluate the effect of software. This evaluation must be done in order to be sure that the communication software will support the line control method that is best from other standpoints. Also, the software evaluation will fix the amount of overhead (nonproductive tasks) that will be imposed on the system because of protocols, line control characters, etc. This decision must be factored into the design, and the central computer load calculated and verified against its capacity.

At this point, the line control method, type of line (low speed, voice grade, etc.), location of multiplexing centers, stations, and even the central computer may have to be reconsidered. The designer must take into consideration the computer that will be running the system, the front-end communication processor that will be controlling the data communication portion of the system, the multiplexers, concentrators, modems, and the various terminals that will be utilized at each station by the user personnel. Also, feasible backup methods must be evaluated. These may include spare equipment at critical points, spare circuits, switching from a private leased line to a dial-up line, or a manual backup using voice communications to conduct the daily business.

Develop Cost Comparisons (step 8)

This is the eighth step of the systems approach to design. Its purpose is to provide a basis for a rational cost decision among the alternative configurations. The major factors in overall cost are:

- Total mileage of each link
- Cost-per-mile of the type of service chosen
- Line termination charges (service terminals)

The total mileage can be roughly determined by referring to maps or atlases. Figure 8-7, Air Distance Table, is included as a convenient reference to help the designer to determine approximate mileages between various major city pairs in the United States. However, interstate and intrastate mileages are determined, for purposes of telephone company billing, by using a system of vertical and horizontal coordinates. The vertical and horizontal coordinate technique uses the principles of analytic geometry to calculate the distance between two points.

Utilizing the vertical and horizontal coordinate technique, the mileage between Tulsa and San Francisco is calculated to be 1,459 miles (see Figure 8-8). Note that the air distance table (Figure 8-7) shows this mileage as 1,461 miles. Federal Communications Commission Tariff Number 264 gives the vertical and horizontal coordinates for the cities in the United States. Coordinates for an abbreviated list of major cities follow:

	Albuquerque, N. Mex	Amarillo, Tex.	Atlanta, Ga.	Billings, Mont.	Birmingham, Ala.	Boston, Mass.	Buffalo, N.Y.	Burlington, Vt.	Charleston, S.C.	Charlotte, N.C.	Cheyenne, Wyo.	Chicago, Ill.	Cincinnati, Ohio	Cleveland, Ohio	Dallas, Tex.	Denver, Colo.	Des Moines, Iowa	Detroit, Mich.	El Paso, Tex.	Fargo, N. Dak.	Houston, Tex.	Indianapolis, Ind.	Jacksonville, Fla.
Albuquerque, N. Mex		273	1272	744	1138	1972	1580	1878	1539	1457	429	1129	1251	1421	588	334	837	1364	229	961	754	1169	1488
Amarillo, Tex.	273		999	809	866	1722	1338	1640	1266	1185	440	894	992	1173	334	358	626	1124	358	847	533	915	1219
Atlanta, Ga.	1272	999		1519	140	937	697	951	267	227	1229	587	369	554	721	1212	739	597	1291	1114	701	426	285
Billings, Mont.	744	809	1519		1425	1861	1473	1713	1761	1617	370	1073	1304	1369	1092	453	798	1283	973	565	1315	1204	1796
Birmingham, Ala.	1138	866	140	1425		1052	776	1049	402	361	1119	578	406	618	581	1095	670	641	1152	1060	567	433	374
Boston, Mass.	1972	1722	937	1861	1052		400	182	820	721	1735	851	740	551	1551	1769	1159	613	2072	1300	1605	807	1017
Buffalo, N.Y.	1580	1338	697	1473	776	400		304	699	538	1335	454	393	173	1198	1370	760	216	1692	919	1286	435	879
Burlington, Vt.	1878	1640	951	1713	1049	182	304		884	755	1612	749	690	476	1501	1654	1049	516	1995	1149	1580	739	1079
Charleston, S.C.	1539	1266	267	1761	402	820	699	884		177	1486	757	506	609	981	1474	967	681	1552	1317	936	594	197
Charlotte, N.C.	1457	1185	227	1617	361	721	538	755	177		1362	587	335	435	930	1358	819	504	1496	1153	927	428	341
Cheyenne, Wyo.	429	440	1229	370	1119	1735	1335	1612	1486	1362		891	1082	1199	726	96	583	1125	653	563	947	986	1493
Chicago, Ill.	1129	894	587	1073	578	851	454	749	757	587	891		252	308	803	920	309	238	1252	569	940	165	863
Cincinnati, Ohio	1251	992	369	1304	406	740	393	690	506	335	1082	252		222	814	1094	510	235	1335	820	892	100	626
Cleveland, Ohio	1421	1173	554	1369	618	551	173	476	609	435	1199	308	222		1025	1227	617	90	1521	835	1114	263	770
Dallas, Tex.	588	334	721	1092	581	1551	1198	1501	981	930	726	803	814	1025		663	632	999	572	972	225	763	908
Denver, Colo.	334	358	1212	453	1095	1769	1370	1654	1474	1358	96	920	1094	1227	663		610	1156	557	642	879	1000	1467
Des Moines, Iowa	837	626	739	798	670	1159	760	1049	967	819	583	309	510	617	632	610		546	983	397	821	411	1023
Detroit, Mich.	1364	1124	596	1283	641	516	216	516	681	504	1125	238	235	90	999	1156	546		1479	745	1105	240	831
El Paso, Tex.	229	358	1291	973	1152	2072	1692	1995	1552	1496	653	1252	1335	1525	572	557	983	1479		1163	676	1264	1473
Fargo, N. Dak.	961	847	1114	565	1060	1300	919	1149	1317	1153	563	569	820	835	972	642	397	745	1163		1183	725	1399
Houston, Tex.	754	533	701	1315	567	1605	1286	1580	936	927	947	940	892	1114	225	879	821	1105	676	1183		865	821
Indianapolis, Ind.	1169	915	426	1204	433	807	435	739	594	428	986	165	100	263	763	1000	411	240	1264	725	865		699
Jacksonville, Fla.	1488	1219	285	1796	374	1017	879	1079	197	341	1493	863	626	770	908	1467	1023	831	1473	1399	821	699	
Kansas City, Mo.	720	481	676	846	579	1251	861	1161	928	803	560	414	541	700	451	558	180	645	839	549	644	453	950
Knoxville, Tenn.	1280	1009	155	1447	235	818	548	815	316	180	1183	454	219	400	767	1178	651	442	1326	1004	790	290	410
Little Rock, Ark.	816	543	456	1143	325	1259	913	1214	723	649	813	552	524	740	293	780	478	723	847	869	388	483	690
Los Angeles, Calif.	664	937	1936	959	1802	2596	2198	2485	2203	2119	882	1745	1897	2049	1240	831	1438	1983	701	1427	1374	1809	2147
Louisville, Ky.	1178	915	319	1275	331	826	483	780	500	343	1033	269	90	311	726	1038	476	316	1254	818	802	110	597
Memphis, Tenn.	939	667	337	1213	217	1137	803	1100	604	521	902	482	410	630	420	879	485	623	976	882	484	384	590
Miami, Fla.	1598	1441	604	2085	665	1255	1181	1347	482	652	1763	1188	952	1087	1111	1726	1333	1152	1643	1716	968	1024	326
Minneapolis, Minn.	983	812	907	742	862	1123	731	985	1104	939	642	355	605	630	862	700	235	543	1157	214	1056	511	1191
Nashville, Tenn.	1119	848	214	1309	182	943	627	916	455	340	1032	397	238	459	617	1023	525	470	1169	902	665	251	499
New Orleans, La.	1029	776	424	1479	312	1359	1086	1361	630	649	1131	833	706	924	443	1082	827	939	983	1222	318	712	504
New York, N.Y.	1815	1560	748	1760	864	188	292	260	641	533	1604	713	570	405	1374	1631	1022	482	1905	1210	1420	646	838
Omaha, Nebr.	721	526	817	703	732	1282	883	1171	1058	918	463	432	622	739	586	488	123	669	878	390	794	525	1098
Philadelphia, Pa.	1753	1494	666	1727	783	271	279	328	562	451	1556	666	503	360	1299	1579	973	443	1836	1184	1341	585	758
Phoenix, Ariz.	330	598	1592	872	1456	2300	1906	2202	1857	1783	663	1453	1581	1749	887	586	1155	1690	346	1225	1017	1499	1794
Pittsburgh, Pa.	1499	1244	521	1479	608	483	178	445	528	362	1298	410	257	115	1070	1320	715	205	1590	949	1137	330	703
Portland, Oreg.	1107	1304	2172	686	2066	2540	2156	2385	2425	2290	947	1758	1985	2055	1633	982	1475	1969	1286	1239	1836	1885	2439
Raleigh, N.C.	1576	1306	356	1698	491	609	490	605	220	130	1461	642	396	428	1057	1463	902	510	1621	1210	1056	495	414
St. Louis, Mo.	942	685	467	1057	400	1038	662	966	704	568	795	262	309	492	547	796	273	455	1034	660	679	231	751
Salt Lake City, Utah	484	668	1583	387	1466	2099	1699	1969	1845	1727	371	1260	1453	1568	999	371	953	1492	689	863	1200	1356	1837
San Antonio, Tex.	617	444	882	1252	744	1766	1430	1729	1122	1105	882	1051	1039	1256	252	802	882	1238	503	1207	189	999	1011
San Francisco, Calif.	896	1157	2139	904	2013	2699	2300	2568	2405	2301	967	1858	2043	2166	1483	949	1550	2091	995	1446	1645	1949	2374
Seattle, Wash.	1184	1359	2182	668	2082	2493	2117	2333	2428	2285	973	1737	1972	2026	1681	1021	1467	1938	1376	1197	1891	1872	2455
Spokane, Wash.	1030	1176	1961	443	1865	2266	1888	2108	2204	2059	768	1508	1744	1796	1489	826	1240	1709	1239	969	1704	1644	2237
Syracuse, N.Y.	1718	1461	781	1600	875	264	138	177	738	595	1472	592	514	303	1326	1508	898	354	1828	1042	1403	567	928
Tulsa, Okla.	604	335	678	930	552	1398	1023	1327	945	853	588	598	661	853	236	550	396	613	674	741	432	591	921
Washington, D.C.	1653	1391	543	1669	661	393	292	432	453	330	1477	597	404	306	1185	1494	896	396	1728	1140	1220	494	647
Wichita, Kans.	549	304	776	801	658	1424	1036	1337	1039	933	465	591	702	873	340	437	334	821	661	634	559	620	1031

Figure 8-7: Air Distance Table

Cities	Vertical	Horizontal
Phoenix	9135	6748
Los Angeles	9213	7878
San Diego	9468	7629
Oakland	8486	8695
San Francisco	8492	8719
Denver	7501	5899
Washington, D.C.	5622	1583
Kansas City, Kansas	7028	4212
Kansas City, Mo.	7027	4203

Kansas City, Mo.	Knoxville, Tenn.	Little Rock, Ark.	Los Angeles, Calif.	Louisville, Ky.	Memphis, Tenn.	Miami, Fla.	Minneapolis, Minn.	Nashville, Tenn.	New Orleans, La.	New York, N.Y.	Omaha, Nebr.	Philadelphia, Pa.	Phoenix, Ariz.	Pittsburgh, Pa.	Portland, Oreg.	Raleigh, N.C.	St. Louis, Mo.	Salt Lake City, Utah	San Antonio, Tex.	San Francisco, Calif.	Seattle, Wash.	Spokane, Wash.	Syracuse, N.Y.	Tulsa, Okla.	Washington, D.C.	Wichita, Kans.
720	1280	816	664	1178	939	1698	983	1119	1029	1815	721	1753	330	1499	1107	1576	942	484	617	896	1184	1030	1718	604	1653	549
481	1009	543	937	915	667	1441	812	848	776	1560	526	1494	598	1244	1304	1306	685	668	444	1157	1359	1176	1475	335	1391	304
676	155	456	1936	319	337	604	907	214	424	748	817	666	1592	521	2172	356	467	1583	882	2139	2182	1961	781	678	543	776
846	1447	1143	959	1275	1213	2085	742	1309	1479	1760	703	1727	872	1479	686	1698	1057	387	1252	904	668	443	1600	930	1669	801
579	235	325	1802	331	217	665	862	182	312	864	732	783	1456	608	2066	491	400	1466	744	2013	2082	1865	875	552	661	658
1251	818	1259	2596	826	1137	1255	1123	943	1359	188	1282	271	2300	483	2540	609	1038	2099	1766	2699	2493	2266	264	1398	393	1424
861	548	913	2198	483	803	1181	731	627	1086	292	883	279	1906	178	2156	490	662	1699	1430	2300	2117	1888	138	1023	292	1036
1161	815	1214	2485	780	1100	1347	985	916	1361	260	1171	328	2202	445	2385	665	966	1969	1729	2568	2333	2108	177	1327	432	1337
928	316	723	2203	500	604	482	1104	455	630	641	1058	562	1857	528	2425	220	704	1845	1122	2405	2428	2204	738	945	453	1039
803	180	649	2119	343	521	652	939	340	649	533	918	451	1783	362	2290	130	568	1727	1105	2301	2285	2059	595	853	330	933
560	1183	813	882	1033	902	1763	642	1032	1131	1604	463	1556	663	1298	947	1461	795	371	882	967	973	768	1472	588	1477	465
414	454	552	1745	269	482	1188	355	397	833	713	432	646	1453	410	1758	642	262	1216	1051	1858	1737	1508	592	598	597	591
541	219	524	1897	90	410	952	605	238	706	570	622	503	1581	257	1985	396	309	1453	1039	2043	1972	1744	514	661	404	702
700	400	740	2049	311	630	1087	630	459	924	405	739	360	1749	115	2055	428	492	1568	1256	2166	1796	1953	303	853	306	873
451	767	293	1240	726	420	1111	862	617	443	1374	586	1299	887	1070	1633	1057	547	999	252	1483	1681	1489	1326	236	1185	340
558	1178	780	831	1038	879	1726	700	1023	1082	1631	488	1579	586	1320	982	1463	796	371	802	949	1021	826	1508	550	1494	437
180	651	478	1438	476	485	1333	235	525	827	1022	123	973	1155	711	2156	902	273	953	882	1550	1467	1240	898	396	896	334
645	442	723	1983	316	623	1152	543	470	939	482	669	443	1690	205	1969	510	455	1492	1238	2091	1938	1709	354	813	396	821
839	1323	847	701	1254	976	1643	1157	1169	983	1905	878	1836	346	1590	1186	1621	1034	689	503	995	1376	1239	1828	674	1728	661
549	1004	869	1427	818	882	1716	214	902	1222	1210	390	1184	1225	949	1239	1210	660	863	1207	1446	1197	969	1042	741	1140	634
644	790	388	1374	803	484	968	1056	665	318	1420	794	1341	1017	1137	1836	1056	679	1200	189	1645	1891	1704	1403	442	1220	559
453	290	483	1809	107	384	1024	511	251	712	646	525	585	1499	330	1885	495	231	1356	999	1949	1872	1646	567	591	494	620
950	410	690	2147	594	590	326	1191	499	504	838	1098	758	1794	703	2439	414	751	1837	1011	2374	2455	2237	928	921	647	1031
	624	325	1356	480	369	1241	413	473	680	1097	166	1038	1049	781	1497	905	238	925	702	1506	1506	1287	998	216	945	177
624		479	1941	188	350	736	792	161	547	632	745	552	1607	375	2115	296	392	1547	959	2121	2114	1890	641	676	923	753
1287	1890	1573	940	1717	1650	2520	1166	1752	1898	2179	1146	2151	1019	1908	290	2139	1500	550	1614	727	229		2010	1353	2100	1227
989	641	1038	2336	603	923	1212	861	739	1187	194	1021	220	2044	268	2281	519	796	1835	1553	2435	2238	2010		1157	290	1173
216	676	231	1266	582	341	1176	626	515	548	1231	352	1163	932	917	1531	972	361	917	486	1461	1560	1353	1157		1058	130
945	430	892	2300	476	765	923	934	569	966	205	1014	123	1983	192	2354	233	712	1848	1388	2442	2359	2100	290	1058		1006
177	753	348	1197	633	442	1297	546	594	677	1266	257	1204	879	950	1411	1044	394	808	573	1369	1437	1227	1173	130	1106	

Dallas	8436	4034
Little Rock	7721	3451
Tulsa	7707	4173
Oklahoma City	7947	4373
Amarillo	8266	5076
Omaha	6687	4595
El Paso	9231	5665
Salt Lake City	7576	7065
Portland	6799	8914
Spokane	6247	8180
Seattle	6336	8896

$$\text{Distance} = \sqrt{\frac{(V_1 - V_2)^2 + (H_1 - H_2)^2}{10}}$$

Location	Vertical	Horizontal
Tulsa	7707	4173
San Francisco	8492	8719

$$D = \sqrt{\frac{(7707 - 8492)^2 + (4173 - 8719)^2}{10}}$$

$D = 1459$ miles

Figure 8-8: Mileage Calculation Using the Coordinate Technique Between Tulsa and San Francisco

Using this method, the telephone company calculates the IXC mileage between the telephone company's central office that services the customer premises at one location and the telephone company central office that services the customer's premises at the other location. Figure 8-9 shows the interexchange channel and the local loops that run from the telephone company central offices to the customer premises. The service terminals in Figure 8-9 are plugs or junction boxes that the telephone company installs at the customer's premises to connect to the modems (whether supplied by the telephone company or the customer). What the customer gets from the telephone company is a complete channel end-to-end or "plug-to-plug."

The calculations of network cost also depend on the type of service involved; they are, indeed, specified by the tariff itself. Suppose, for example, that prior design analysis has determined that, for a network of three links connecting New York to respectively Chicago, Pittsburgh, and Washington, D.C., Series 1005 lines will suffice. The costs are calculated as shown in Figure 8-10.

Calculating the network link cost for voice grade (Series 2000/3000) lines for the same network, New York to respectively Chicago, Pittsburgh, and Washington D.C. is shown in Figure 8-11.

The network calculator (see Figure 8-12) can be used to accumulate a series of network costs on a link-by-link basis. It can be used for either Series 1000 or 2000/3000 as well as wideband circuits.

At this point, the system designer may wish to make many iterations back through Step 7 to include rethinking of line control methods, network configurations, and other variables.

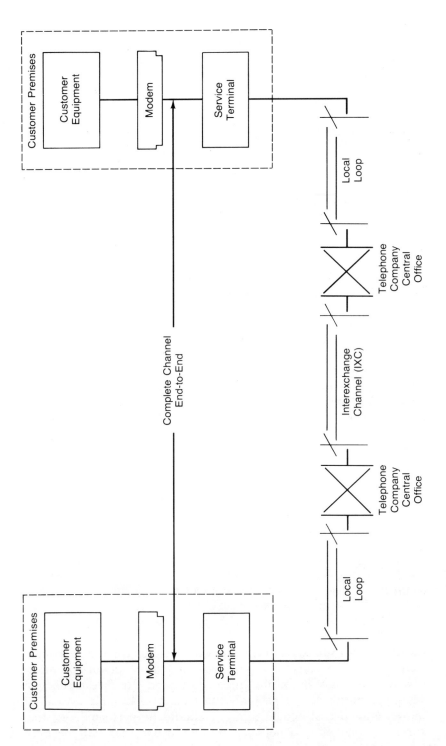

Figure 8-9: Interexchange Channel Mileage (IXC)

Link	Miles		Dollars/Mile		Subtotals	Totals
New York–Chicago	100	×	$1.25	=	$125.00	
(713 miles)	150	×	1.00	=	150.00	
	250	×	0.60	=	150.00	
	213	×	0.40	=	85.20	
						$ 510.20
New York–Pittsburgh	100	×	$1.25	=	$125.00	
(317 miles)	150	×	1.00	=	150.00	
	67	×	0.60	=	40.25	
						$ 315.20
New York–Washington	100	×	$1.25	=	$125.00	
(205 miles)	105	×	1.00	=	105.00	
						$ 230.00
					Total cost*	$1,055.40

*IXC cost only

Figure 8-10: Line Costs (Type 1005—Low-Speed) for
Point-to-Point Links

Link	Miles		Dollars/Mile		Subtotals	Totals
New York–Chicago	100			=	$175.20	
(713 miles)	613	×	$0.66	=	404.58	
						$ 579.78
New York–Pittsburgh	100			=	$175.20	
(317 miles)	217	×	$0.66	=	143.22	
						$ 318.42
New York–Washington	100			=	$175.20	
(205 miles)	105	×	$0.66	=	69.30	
						$ 244.50
					Total cost*	$1,142.70

*IXC cost only

Figure 8-11: Line Costs (Series 2000/3000—Voice
Grade) for Point-to-Point Links

Figure 8-13 lists various factors to consider when performing a cost-benefit analysis for a data communication network. A quick cost-benefit analysis for the direct costs and the direct benefits should be performed on each alternative network configuration. A more detailed cost-benefit analysis should be performed on the surviving one or two network configurations that will be pre-

Link	Circuit Cost	Station Terminals	Modems	Multi-plexers	Terminals	Link Total Cost
San Fran-cisco/ Bakers-field	$278.82	50.00	60.00*	None	180.00†	$568.82
Bakers-field/ Los Angeles	$164.00	·25.00	30.00	None	90.00	$309.00

*Assume $30.00 per month
†Assume $90.00 per month

Grand Total $877.82 per month

Figure 8-12:   Network Calculator (The Data Are from Figure 7-18)

sented to management. The latter should include comparative details needed by management to make its decision.

Selling the System (step 9)

This step is overlooked or underemphasized in many systems projects, often with serious consequences. We believe that the system designer's product is not a paper design, but is, instead, a satisfied user. Selling the system is a step toward making the user satisfied.

COSTS

Direct costs:
—Computer equipment
—Communications equipment
—Common carrier line charges
—Software
—Operations personnel costs
—File conversion costs
—Facilities costs (space, power, air-conditioning, storage space, offices, etc.
—Spare parts costs
—Hardware maintenance costs
—Software maintenance costs
—Interaction with vendor and/or development group
—Development and performance of acceptance test procedures and parallel operation
—Development of documentation
—Costs for back-up of system in case of failure
—Costs of manually performing tests during a system outage

Indirect costs:
—Personnel training
—Transformation of operational procedures
—Development of support software
—Disruption of normal activities
—Increased system outage rate during initial operation period
—Increase in the number of vendors (impacts fault detection and correction due to "finger pointing")

BENEFITS

Direct and indirect cost reductions:
—Elimination of clerical personnel and/or manual operations
—Reduction of inventories, manufacturing, sales, operations, and management costs
—Effective cost reduction, e.g., less spoilage or waste, elimination of obsolete materials, and less pilferage
—Distribution of resources across demand for service

Revenue increases:
—Increased sales due to better responsiveness
—Improved services
—Faster processing of operations

Intangible benefits:
—Smoothing of operational flows
—Reduced volume of paper produced and handled
—Rise in level of service quality and performance
—Expansion capability
—Improved decision process by providing faster access to information
—Ability to meet the competition
—Future cost avoidance
—Positive effect on other classes of investments or resources such as better utilization of money, more efficient use of floor space or personnel, etc.
—Improved employee morale

Figure 8-13:   Cost-Benefit Analysis Factors

To sell the system to management requires thorough, lucid documentation that supports the logic of the system. Management will want to see:

- A clear, brief statement of the problem and the consequences of continuing without a solution.

- A clear, brief statement of the proposed solution, described in functional terms.

- A cost-benefit analysis of the alternative solutions, which supports the selection of the proposed solution.

- A list of problems that are anticipated.

- An implementation schedule and costs.

Larger projects may require that the recommendations be put into formal report or presentation form. Prepare the report and presentation well; do not allow poor planning at this stage to invalidate a good systems design job. If you are unfamiliar with formal report or presentation methods, obtain guidance from books* or from associates who have experience.

During the presentation, try to deduce the objections (or questions) management may think of and respond to them in advance in your documentation. Strive for lucidity and make sure that the focus of the report or presentation is always on answering the question "Why is adopting this system to the overall benefit of the organization?"

Implementation, Follow-up, and Reevaluation (step 10)

Just because management "buys" the design does not mean the designer's job is finished. The designer has an obligation to monitor all the activities of implementation, and to ensure that all aspects of the design are brought into being professionally and faithfully. In some instances, the designer is the implementation manager. But whatever the case, it is important to prevent design decisions from being circumvented or even reversed by the ill-considered actions of implementers who may not fully understand the rationale of the system design. The chances are good that implementation failures will be blamed on "poor design," so maintaining surveillance over the implementation processes and schedules is important. The designer must be especially careful not to accept substandard services or facilities from vendors. The designer must plan and conduct deliberate, formal tests before acceptance.

After the system is installed and running, it is time to evaluate its performance. The designer should examine the following areas and be satisfied they are all performing properly:

- Terminal equipment
- Modems, multiplexers, concentrators, etc.
- Communication lines
- Computers/front-end processes
- Software

---

*John M. FitzGerald and Ardra F. FitzGerald, *Fundamentals of Systems Analysis* (New York: John Wiley & Sons, Inc., 1973), Chapter 11.

- Users
- Training
- Documentation
- Maintenance
- Network management

The criteria for performance are those established in Step 6—Define the system requirements. Look for differences between the estimated traffic volume and the actual traffic volume. Discover the reasons for these differences. This is especially important when the actual volume is lower than was estimated—it may be symptomatic of something wrong in the system, e.g., unreliability, poor training. In general, adopt the viewpoint of a physician doing an annual physical examination: compare system characteristics against requirements, look for abnormalities, and investigate, explain, and prescribe treatment for any abnormalities you find.

Finally, recognize that the follow-up and reevaluation are not just one-time processes. They should be repeated at regular intervals throughout the life of the system.

Questions—Chapter 8

The following questions cover a range of material. They are complemented by a series of complete design problems that are contained in the instructor's workbook. Together, you may consider these problems as the measure of whether you are competent to work in this field in business, industry, or government. This is a demanding field, and a field in which competent people are in demand.

True or False

1.   The systems approach is another name for the trial and error approach.

2.   The first basic design step is to prepare an outline of the approach and methodology that will be utilized in studying and designing the proposed system.

3.   Poor planning is the reason for many systems failures.

4. Quantitative performance measurements are preferable to qualitative performance measurements when designing the system's requirements.

5. One of the substeps of designing the proposed data communication system would be to determine message volumes.

6. A proposed system is likely to have little impact outside the mainstream application area.

7. In determining message volumes, it is sufficient to concentrate on average daily volumes.

8. A character count of a business message is a good estimate of the number of characters that will be transmitted to a video terminal.

9. Follow-up and reevaluation should be repeated at regular intervals.

10. Should a multiplexer fail, it is likely that several terminals would be affected.

Fill-in

1. The most critical basic step of designing a communication system is to ———— ———— ———— ————.

2. As part of the design of the proposed data communication system, one should develop ———— configurations.

3. An understanding of the ———— ———— will provide a benchmark for measurement and a point from which to start system design efforts.

4. The last of the basic design steps discussed is ————, ———— and, ————.

5. There are ———— basic steps to system design presented in the text.

6. The ———— ———— results when the system users utilize the system to a greater extent than was anticipated because the system is found to be more efficient, available, and quick.

7. The systems designer may have to plan a greater throughput so that ———— ———— ———— can get through the network without inordinate delays.

8. The systems designer should probably consider a ⎯⎯ to ⎯⎯ % contingency factor for the retransmission of messages that were received in error.

9. Interstate and intrastate mileages are determined, for purposes of actual telephone company billings, by using a set of ⎯⎯ and ⎯⎯ ⎯⎯.

10. The major factors in overall cost of a communication network are total mileage of each link, cost-per-mile of the type of service chosen, and ⎯⎯ ⎯⎯ ⎯⎯.

Multiple Choice

1. The design plan should include:

    a) Sources of information to be used
    b) Types of information to be collected
    c) Schedules of all activities
    d) All the above
    e) None of the above

2. Among the activities required to design the proposed data communication system would be:

    a) Analyze the types of messages/transactions
    b) Calculate the network cost
    c) Establish the line loading
    d) A and C above
    e) All the above

3. What is the correct ordering of the following basic design steps?

    a) Gather background information, understand the existing system, prepare an analysis and design plan
    b) Gather background information, prepare an analysis and design plan, understand the current system
    c) Prepare an analysis and design plan, gather background information, understand the current system
    d) Prepare an analysis and design plan, understand the current system, gather background information
    e) Understand the current system, gather background information, prepare an analysis and design plan

4. It is suggested that each message should be identified by a short title as part of which of the following design steps?

    a) Defining the systems requirements

b) Designing the proposed data communication system
c) Gathering background information
d) Understanding the current system
e) None of the above

5. To establish the line load, which of the following should be considered?

a) Computer growth factor
b) Efficiency of code utilized
c) Transmission line errors
d) B and C above
e) All the above

Short Answer

1. The following is an excerpt from last month's progress memo from the data communication analyst Jones to Mr. Smith, his manager.

Mr. Allen, VP Marketing, called me to ask if anything could be done to improve the order entry system. I met with him and the manager of marketing administration, Mrs. Johnson, and listened to their problems. We agreed on a short written definition of the problem. I then wrote an outline of the approach to the problem, and determined that marketing field offices and salesmen, marketing headquarters, manufacturing, and distribution will be affected by any changes to the present system. Mr. Allen turned down my request to visit a typical field office because he feels Mrs. Johnson knows enough about their operation to fill me in. I met with the manufacturing planner, Mr. Williams, and the head of distribution, Mr. Thomas. I obtained a general understanding of the current order entry system from Johnson, Williams, and Thomas. The design of the new system is underway now, and will be finished soon. I will then prepare a cost estimate of the system design and will present both to you in my next month's report.

Play the role of Mr. Smith and write a memo to Jones, commenting on his report. Be critical and try to determine any areas where Jones may not have done all that he should have done.

2. Critique the following excerpts from a system requirements document:

- "The system shall be easy to operate"

- "The system shall have a mean time between failures of at least 1,000 hours"

- "The system shall transmit in half duplex mode at 2,400 bps"
- "The system shall transmit at least 1,000 messages per hour"

3.   The busiest link in a system carries 500,000 10-bit characters per 12-hour day. Allowing for a peak load equal to three times the average, for a 50% growth over the system life, and for a 10% error/retransmission factor, what is the minimum line speed in bits per second?

4.   Using the data from the table in the section "Develop Cost Comparisons (Step 8)" and the formula from Figure 8-8, calculate the distance from San Francisco to Denver and from Kansas City, Kansas, to Kansas City, Missouri.

5.   In which step of the system design process would your plan start to identify the sources of information to be used, the types of information to be collected, the analyses to be performed, the schedules of the various activities, and the definition of the results to be produced.

6.   Using Figure 8-4: Network Link Total Traffic, fill in some hypothetical figures for the various headings across the top. These figures will be used in Question 7 below.

7.   Convert the characters per day to bits per second as shown in Figure 8-5 by utilizing those figures that you have developed in answering Question 6 above.

8.   What are some of the items that utilize transmission time, but do not transmit business data and for which the designer must account?

9.   Give examples of what factors might cause peak loads and discuss how the system designer would take them into account when designing a data communication system.

10.   All the network design problems are in the instructor's manual because of their extensive size and the maps and tables required. However, the following pages contain a simple problem that has been worked out as an example. Review this problem and then answer Questions a, b, and c.

    a. How many modems, channel terminals, service terminals, and terminals are required at LA, SF, NY, and NO?

b. Can this firm move from a type 1006 line to a type 1005 line without sacrificing throughput or contingency factors? Explain your answer.

c. Can the line mileages be reduced and still use a type 1006 line? Explain your answers.

A Proposed Data Communication Network
for Fitson, Inc.

The home office, located in Los Angeles, Ca. will have a terminal to inquire about and maintain inventory data. The following branch offices will also have a terminal: San Francisco, New Orleans, Los Angeles, and New York City.

Assumptions:

- (1) All four stations (LA, SF, NO, NY) have the same number of messages and message lengths.
- (2) There are no peak periods or seasonal fluctuations.
- (3) The LA computer responds to all messages.
- (4) If a station is unable to transmit all its data by the end of the work day, overtime will be authorized for personnel to transmit the data after working hours.
- (5) The LA terminal is hard-wired (no phone lines).
- (6) Hours available for transmission between the home office and the branches (local time) are:

	California	Louisiana	New York
	8 AM — 5 PM	10 AM — 5 PM	11 AM — 5 PM
Hours available	9	7	6

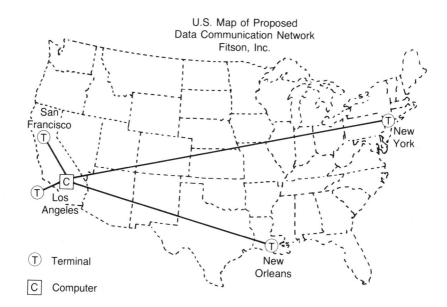

U.S. Map of Proposed
Data Communication Network
Fitson, Inc.

San Francisco

New York

Los Angeles

New Orleans

(T) Terminal

[C] Computer

Line Requirement

1,900	Average messages per day per station (in and outbound)*
× 70	average characters per message (inbound equals outbound length)†
133,000	characters per day per station
6,650	characters—5% for turnpike effect
13,300	characters—10% for growth
2,660	characters—2% allowance for error rate
155,610	characters—per day per station—corrected
× 11	11-bit code—asynchronous transmission—8 plus 3
1,711,710	bits per day per station
÷ 6	six hour work day—shortest work day (NY)
285,285	bits per hour per station
÷ 60	convert to bits per minute
4,755	bits per minute
÷ 60	convert to bits per second
80	bits per second

Other contingency factors may also be considered.

Requires a line that will handle 80 bps

---

*Peaks may change this calculation; each station usually has a different number of messages but in this case all are assumed equal.

†In and outbound messages usually have a different number of characters; peak message lengths may change this calculation.

Calculations of Line Rates—for a Half Duplex Type 1006 Line

Cities	Miles	Miles		Rate		Cost	Total Cost
LA — SF	347	100	×	$1.55	=	$155.00	
		150	×	1.25	=	187.50	
		97	×	0.80	=	77.60	
							$420.10
LA — NO	1,673	100	×	$1.55	=	$155.00	
		150	×	1.25	=	187.50	
		250	×	0.80	=	200.00	
		500	×	0.50	=	250.00	
		673	×	0.30	=	201.90	
							$994.40
LA — NY	2,451	1000			=	$792.50	
		1451	×	0.30	=	435.30	
							$1,227.80
							$2,642.30

Glossary

Analog Pertaining to representation by means of continuously variable physical quantities, such as varying voltages, frequencies, etc.

Analog channel A channel on which the information transmitted can take any value between the limits defined by the channel.

Asynchronous
1. Having a variable time interval between successive bits, characters, or events. In asynchronous data transmission, each character is individually synchronized, usually by using start and stop bits.
2. Descriptive of the transmission method, or the terminal equipment employed, which demands timing information be included in the transmitted character.

Asynchronous transmission
See Transmission, Asynchronous

Attenuation The difference between transmitted and received power due to loss through equipment, lines, or other transmission devices.

Auto-answer (also Automatic answer) An equipment feature that allows a station to automatically respond to a call that it receives over a network.

Automatic calling unit A unit that will generate dial pulses or tones to a telephone network in response to signals from a computer or business machine.

Bandwidth The difference between the highest and lowest frequencies in a band, such as 3000 cycles bandwidth in a voice grade line (300–3,300 cycles).

Baud A unit of signaling speed equal to the number of discrete conditions or signal events per second.

Baudot code
See Code, Baudot

BCD
See Code, BCD

Bit
1. An abbreviation of the words "binary digit."
2. A single character in a binary number.
3. A single pulse in a group of pulses.
4. A unit of information capacity of a storage device.

Block A group of bits, or characters, transmitted as a unit. An encoding procedure is generally applied to the group of bits or characters for error control purposes.

bps Bits per second.

Byte A binary character string operated on as a unit and usually shorter than a computer word. Unless otherwise specified, a byte normally contains 8 bits.

Carrier A continuous frequency capable of being modulated with a second information-carrying signal.

Carrier, Common Organizations licensed and regulated by the United States Federal Communications Commission or the various state public utility commissions which supply communication services to users at published prices.

CCSA
See Common Control Switching Arrangement

Central computer
See Computer, Central

Centrex A telephone PABX equipment service that allows dialing within the system, direct inward dialing, automatic identification of outward dialing, and can be used to limit outward long distance dialing.

Channel A path for transmission of electromagnetic signals. Synonym for line and link. Compare with circuit.

Character A member of a set of elements upon which agreement has been reached and that is used for the organization, control, or representation of data. Characters may be letters, digits, punctuation marks, or other symbols.

Checking, Echo A method of checking the accuracy of transmission data in which the received data are returned to the sending end for comparison with the original data.

Checking, Loop Synonym for echo checking.

Checking, Parity A check that tests whether the number of ones (or zeroes) in an array of binary digits is odd (or even).

Checking, Polynomial A checking method using polynomial functions of the data transmitted to test for changes in data in transmission.

Circuit A means of two way communication between two data terminal installations. Compare with channel, line, link.

Code A transformation or representation of information in a different form according to some set of preestablished conventions.

Code, Baudot A data transmission code in which 5 bits are used to represent a character.

Code, BCD A code for decimal notation in which the decimal digits are represented by a binary code group, generally in 8, 4, 2, 1 weighted notation.

Code, Constant Ratio A code in which the ratio of 1's and 0's in each character is maintained constant.

Code, EBCDIC   An acronym for Extended Binary Coded Decimal Interchange Code. A standard code consisting of a character set of 8-bit characters used for information representation and interchange among data processing and communication systems. Very common in IBM equipment.

Code, USASCII   An acronym for United States of America Standard Code for Information Interchange, an 8-bit code used for many data transmission applications and treated as standard among most equipment manufacturers.

Common Control Switching Arrangement (CCSA).   A dedicated switched network leased by a user to handle communication requirements among various locations.

Communication processor, front-end   An auxiliary processor that is placed between a computer central processing unit and transmission facilities. This device normally handles housekeeping functions such as management of lines, translation of codes, etc. which would otherwise interfere with efficient operation of the central processing unit. Synonym for front-end computer.

Computer, Central   In data transmission, the computer that lies at the center of the network and generally does the basic centralized functions for which the network was designed. Synonym for host computer.

Computer, Front-End
See Communication processor, front-end

Concentrator   Equipment designed to improve the efficiency of data or voice transmission by allowing terminals or lines to compete for and share transmission channels.

Conditioning   The electrical balancing of a leased voice grade channel to provide improved electrical characteristics required for high-speed transmission. The primary result is the equalizing of the attenuation distortion between the higher and lower frequencies so all bits reach their destination at the same time.

Constant Ratio
See Code, Constant Ratio

CPU   Central Processing Unit

Cross talk noise
See Noise, Cross talk

CRT Terminal
See Terminal, CRT

Data acquisition   The process of identifying, isolating, and gathering source data to be processed in a usable form.

Data communications
  1. The movement of encoded information by means of electrical transmission systems.
  2. The transmission of data from one point to another.

Dibit   A group of two bits. In four-phase modulation, each possible dibit is encoded as one of four unique carrier phase shifts. The four possible states for a dibit are 00, 01, 10, 11.

Digital signal A discrete or discontinuous signal. Pertains to data in the form of digits such as 0's and 1's.

Direct distance dialing A method of making toll telephone calls on the public network without the aid of an operator.

Distortion, Delay Distortion occurring when the inherent delay of circuit or system is not constant over the frequency range required for transmission.

Distributed functions Moving processing functions from a central computer to remotely located minicomputers or intelligent terminals.

EBCDIC
See Code, EBCDIC

Echo checking
See Checking, Echo

Echo noise
See Noise, Echo

Efficiency, Transmission The ratio, usually expressed in terms of percent, derived by dividing the useful information transmitted by the total information transmitted.

Facilities The elements of the telephone plant that provide a complete connection, exclusive of the customer's equipment.

FDX
See Full duplex

Foreign Exchange (FX) Service that connects a customer's telephone to a telephone company's central office not normally servicing the customer's location.

Frequency Shift Keying (FSK) A method of transmission whereby the carrier frequency is shifted up and down from a mean value in accordance with the binary signal; one frequency represents a binary one, while the other represents a binary zero.

Front-end communication processor
See Communication Processor

FSK
See Frequency Shift Keying

Full duplex (FDX) A circuit that permits transmission of a signal in two directions simultaneously (sometimes called Duplex).

FX
See Foreign Exchange

Gaussian noise
See Noise, Gaussian

Half Duplex (HDX) A circuit that permits transmission of a signal in two directions, but not at the same time. Contrast with full duplex.

Hardware A generic, somewhat slang term used to include all equipment, both computer and communications, contained in a system. Contrast with software.

Hard-wired A slang, but descriptive, term used to indicate equipment or procedures that are fixed in design and usually inflexible in use.

HDX
    See Half duplex

Hertz (Hz)
    Same as cycles per second; e.g., 3,000 Hertz is 3,000 cycles per second.

Hot Line   A service offered by Western Union that provides direct connection between customers in various cities using a dedicated line network.

Impulse noise
    See Noise, impulse

Interexchange Channel (IXC)   A channel between exchanges.

Intermodulation noise
    See Noise, intermodulation

Isochronous transmission
    See Transmission, Isochronous

IXC
    See Interexchange channel

Jitter
    1. A temporary waning defect created by an instability in the transmission signal.
    2. A shift in the time or phase position of individual pulses, causing difficulty in synchronization or detection.

Leased line
    See Line, Leased

Line   A channel or link.

Line control   The collection of operating procedures and control signals by which a data transmission system is controlled.

Line, Leased (or Private)   A line furnished to a subscriber for his exclusive use.

Line loading   The total amount of transmission traffic carried by a line, usually expressed as a percentage of the total theoretical capacity of that line.

Line, Local Loop   A communication line connecting several terminals in the region of a single controller to that controller.

Line protocol   The detailed procedure for the exchange of signals between a source and a sink, designed to accomplish message transmission.

Link   A channel or a line, normally restricted in use to a point-to-point line.

Local loop
    See Loop, Local

Log
    1. A record of everything pertinent to a system function.
    2. A collection of messages that provides a history of message traffic.

Loop-back   The procedure of returning signals through special switching to a test device or center so that the characteristics and operational status of a circuit may be evaluated.

Loop checking
See Checking, Loop

Loop, Local   That part of a communication circuit between the customer's location and the nearest central office.

Low-Speed Lines (Series 1000)   The AT&T designation for lines used for data transmission at rates up to 150 BPS.

Maintainability   The generic term for the properties of equipment or systems that ensure that repairs can be made quickly and simply. Maintainability is usually measured in terms of the mean-time-to-repair, a statistical measure of the average time to repair any system failure.

Message   A communication of information from a source to one or more destinations, usually in code. A message is usually composed of three parts:
1.   A heading, containing a suitable indicator of the beginning of the message together with some of the following information: source, destination, date, time, routing.
2.   A body containing information to be communicated.
3.   An ending containing a suitable indicator of the end of the message.

Message switching   A data communication operation in which messages are switched from one terminal location to another. Both lines must be free.
See Store-and-forward.

Metered Service (WATS)   This service, called Wide Area Telecommunications Service by AT&T, is a combination of offerings of bulk long distance service under various terms involving flat-rate charges per hour of usage.

Modem   A contraction of the words "modulator-demodulator." A modem is a device for performing necessary signal transformation between terminal devices and communication lines. They are normally used in pairs, one at either end of the communication line.

Modular   A program or unit of hardware that is discrete and identifiable and is designed for use with other programs or hardware components.

Modulation   Alteration in the characteristic of a carrier signal by impressing an information signal on the carrier.

Modulation, Amplitude   The form of modulation in which the amplitude of the carrier is varied in accordance with the instantaneous value of the modulating signal.

Modulation, Frequency   A form of modulation in which the frequency of the carrier is varied in accordance with the instantaneous value of the modulating signal.

Modulation, Phase   A form of modulation in which the phase of the carrier is varied in accordance with the instantaneous value of the modulating signal.

Modulation, Pulse Code (PCM)   A form of modulation in which the modulating signal is sampled and the sample quantized and coded so that each element of information consists of different kinds or numbers of pulses and spaces.

Multidrop (Multipoint) A line or circuit interconnecting several stations.

Multiplexing The subdivision of a transmission channel into two or more separate channels. This can be achieved by splitting the frequency range of the channel into narrower frequency bands (frequency division multiplexing) or by assigning a given channel successively to several different users at different times (time division multiplexing).

NCP
See Network Control Program

Network A system consisting of a number of terminal points that are able to access one another through a series of communication lines and switching arrangements.

Network Control Program (NCP) The program within the software system for a data processing system which deals with the control of the network. Normally it manages the allocation, use, and diagnosis of performance of all lines in the network and of the availability of the terminals at the ends of the network. NCP is also used as a specific term referring to a component of SNA (Systems Network Architecture)

Noise Unwanted signals originating in a channel.

Noise, Cross talk Noise resulting from the interchange of signals on two adjacent channels.

Noise, Echo On voice-grade lines with improper echo suppression, the "hollow" or echoing characteristic that results.

Noise, Gaussian Noise that is characterized statistically by a Gaussian, or random, distribution.

Noise, Impulse Noise caused by individual impulses on the channel.

Noise, Intermodulation Noise resulting from the intermodulation products of two signals. This is a result of harmonic reinforcements and cancellations of frequencies.

Noise, White Noise that has equal energy at all frequencies.

Office The common designation for any facility in the public switched network at which switching takes place.

Office, Central (or End) The office closest to the subscriber.

Office, Tandem end An office that terminates a tandem trunk.

Office, Toll An office that terminates a toll trunk.

On-Line
1. Pertaining to equipment or devices under the direct control of a central processing unit.
2. Pertaining to a user's ability to interact with a computer.
3. Pertaining to a user's access to a computer via a terminal.

Overhead Computer time used to keep track of or run the system as compared with computer time used to process data.

PABX
See Private Automatic Branch Exchange

Packet switching   The technique used when long messages are subdivided into short packets where the maximum length is fixed. This contrasts with conventional switching systems in which messages are usually transmitted whole, irrespective of length.

Parallel
1. Pertaining to concurrent or simultaneous operation of two or more devices, or the concurrent performance of two or more activities in a single device.
2. Pertaining to the concurrent simultaneous occurrence of two or more related activities in multiple devices or channels.
3. Pertaining to the simultaneous processing of the individual parts of a whole, such as bits of a character and characters of a word using separate facilities with the various parts. Contrast with serial.

Parity checking
See Checking, Parity

PBX
See Private branch exchange

PCM
See Modulation, pulse code

Phase   Pertaining to the relative timing of an alternating signal. Two signals may be identical in amplitude and frequency, but may differ in phase if one signal lags the other by any value not an exact multiple of the frequency.

Point of sale   Denoting data capture at the place and time of sale. Applies to electronic cash registers and more sophisticated data capture equipment of the same type.

Point to Point   Denoting a channel or line that has only two terminals. A link.

Polling   Any procedure that sequentially contacts several terminals in a network.

Polling, Hub-Go-Ahead   Sequential polling, in which the polling device contacts a terminal, that terminal contacts the next terminal, etc.

Polling, Roll call   Polling accomplished from a prespecified list in a fixed sequence, with polling restarted when the list is completed.

Polynomial checking
See Checking, Polynomial

Private Automatic Branch Exchange (PABX)   An automatic PBX. See Centrex

Private Branch Exchange (PBX)   A small telephone exchange installed on a customer's premises to allow internal dialing from station to station within the customer's premises and connection to outgoing and incoming lines.

Process control   Automatic control of a process in which a computer is used for regulation, usually of a continuous operation or process, for example petrochemicals, cement plants, steel plants, etc.

Protocol A procedure for synchronization so that the receiver knows when a bit starts and ends so that it can be sampled, similarly for character synchronization so that the receiver can determine which bit belongs to a character, and similarly for message synchronization so that the receiver can recognize a special character sequence that delineates messages. Typical protocols include the blocking of transmission in the messages, employing start-of-text and end-of-text or other markers, and a positive or negative acknowledgement procedure.

Real time
1. Pertaining to the actual time during which a physical process occurs.
2. Pertaining to the performance of a computation during the actual time that the related physical process occurs, in order that results of the computation can be used in guiding the physical process.

Reliability The characteristic of equipment, software, or systems that relates to the integrity of the system against failure. Reliability is usually measured in terms of mean-time-to-repair, the statistical measure of the interval between successive failures of the system under consideration.

Remote Job Entry (RJE) Submission of jobs (i.e., computer production tasks) through an input unit (terminal) that has access to a computer through data communication facilities.

Reverse-channel A feature of certain modems which allows simultaneous transmission (usually of control or parity information) from the receiver to the transmitter over a half-duplex data link. Generally the reverse channel is a low speed channel.

RJE
See Remote Job Entry

RS232 Interface The interface between a modem and the associated data terminal, as defined by the Electronics Industries Association Standard RS232.

SDLC
See Synchronous Data Link Control

Serial
1. Pertaining to the sequential performance of two or more activities in a single device.
2. Pertaining to the sequential or consecutive occurrence of two or more related activities in a single device or channel.
3. Pertaining to the sequential processing of the individual part of the whole, such as the bits of a character, or the characters of a word using the same facilities for successive parts. Contrast with parallel.

Service Terminal The plug supplied by the Common Carrier into which the modem plugs (Series 1000 channels).

Simplex A circuit capable of transmission in one direction only. Contrast with half duplex, full duplex.

SNA
See Systems Network Architecture

Software A generic, somewhat slang term for a computer program,

sometimes taken to include also documentation and procedures associated with such programs.

Start bit   A bit preceding the group of bits representing a character used
to signal the arrival of the character in asynchronous transmission.

Station
One of the input or output points on a network.

Station Terminal   The plug supplied by the Common Carrier into which
the modem plugs (Series 2000/3000 channels).

Stop bit   A bit used following the group of bits representing a character,
to signal the end of a character in asynchronous transmission.

Store-and-forward   Applied to communication systems in which messages are received at intermediate points and stored. They are then
retransmitted to a further point or to the ultimate destination.
See Message Switching.

Switched network   Any network in which switching is present and
which is used to direct messages from the sender to the ultimate
recipient.

Switched network, Line-switched   A switched network in which switching is accomplished by disconnecting and reconnecting lines in different configurations in order to set up a continuous pathway between the
sender and the recipient.

Switched network, Store-and-forward   A switched network in which the
store-and-forward principle is used to handle transmissions between
senders and recipients.

Synchronous Data Link Control (SDLC)   The term applied by IBM to
the data link control philosophy forming a part of SNA.

Synchronous transmission
See Transmission, Synchronous

System   A collection of people, machines, and methods organized to
accomplish a set of specific functions.

Systems Network Architecture (SNA)   The term applied by IBM to the
conceptual framework used in defining data communication interaction with computer systems.

Tariff   The schedule of rates and regulations pertaining to the services
of a communication common carrier.

Teleprinter   A Teletype or Teletype device, consisting of a keyboard and
a printing device.

Telex   The term applied by Western Union to its international switched
message service.

Telpak   Package services offered by AT&T making available bulk
facilities for a fixed charge. This offering may be deleted.

Terminal, CRT   CRT is the acronym for Cathode Ray Tube, i.e. a video
display device associated with a terminal.
See Terminal, video.

Terminal, Video   A terminal using a video display as a readout device, in contradistinction to a teleprinter, which uses a printer device. Synonym for CRT terminal.

Time-sharing A method of operation (in an on-line system) in which computer facilities are shared by several users for different purposes during the same time period. Although the computer actually services each user in sequence, the high speed of the computer makes it appear that the users are handled simultaneously.

Transmission, Asynchronous Transmission of data in which the entire message does not operate from the same time base. Typically the bits that make up a single character are transmitted synchronously, i.e. at a constant time base. However, there may be arbitrary delays between characters thus making the timing of the characters truly asynchronous.

Transmission, Isochronous Transmission that combines certain characteristics of asynchronous and synchronous transmission. The entire transmission shares a common time base or clock but the intervals between individual characters may be arbitrary multiples of the period to transmit one character.

Transmission, Synchronous In this form of transmission, data are sent continuously against a time base that is shared by transmitting and receiving terminals. If no legitimate data are available to be sent at a given time, "Synch" or "Idle" characters are sent to keep the transmitter and receiver in time synchronization.

Trunk A communication channel between switching devices or central offices.

Trunk, Intertoll A communication channel between two toll offices.

Trunk, Tandem A trunk that connects tandem offices or connects central offices with tandem offices.

Trunk, Toll connecting A communication channel between a toll office and a local central office. A toll trunk is used to connect toll lines to subscriber lines.

Turnaround time The time required to reverse the direction of transmission from send to receive or vice versa on a half-duplex circuit.

TWX The name given by Western Union to its teleprinter exchange service providing real time direct connection between subscribers. TWX service is confined to North America. Contrast with Telex service which is worldwide.

USASCII
See Code, USASCII

Video terminal
See Terminal, Video

Voice-grade (Series 2000/3000) The term applied to channels suitable for transmission of speech and digital or analog data or facsimile, generally with a frequency range of about 300 to 3,000 Hertz.

WATS
See Metered service (WATS)

White noise
See Noise, white

Wideband (Series 8000) The term applied to channels provided by common carriers capable of transferring data at speeds from 19,200 bps up to the 1 million bps region (19.2 kHz to 1,000 kHz).

Word
1. In communications, six characters (five plus a space)
2. In computers, the unit of information transmitted, stored and operated upon at one time.

Bibliography

Selected Books

Auerbach Terminal Equipment Digest. Revised ed. Princeton, N.J.: Auerbach Publishers, Inc.

Bear, Donald. *Telecommunication Traffic Engineering.* Forest Grove, Oregon: International Scholarly Book Services, Inc., 1976. (IEE Telecommunications Series)

Becker, Hal B. *Functional Analysis of Information Networks: A Structured Approach to the Data Communications Environment.* New York: John Wiley and Sons, 1973.

Data Communications Primer. White Plains, N.Y.: IBM Corp., Data Processing Division (IBM Manual No. GC20-1668).

Davenport, William P. *Modern Data Communication: Concepts, Language, and Media.* New York: Hayden Book Co., 1971.

Davies, Donald W. and Derek L. A. Barber. *Communication Networks for Computers.* London: John Wiley and Sons, 1973.

Flood, J. E. ed. *Telecommunications Networks.* Forest Grove, Oregon: International Scholarly Book Services, Inc., 1976. (IEE Telecommunications Series).

Freeman, Roger L. *Telecommunications Transmission Handbook.* New York: John Wiley and Sons, 1975.

Gentle, Edgar C., Jr. *Data Communications in Business: An Introduction.* New York: American Telephone and Telegraph Co., 1965.

Griesinger, Frank Kern. *How to Cut Costs and Improve Service of Your Telephone, Telex, TWX, and other Telecommunications.* New York: McGraw-Hill Book Co., 1974.

Intercity Services Handbook. Latest Ed. New York: American Telephone and Telegraph Co., Long Lines Marketing Department, 110 Belmont Drive (Room B54), Somerset, New Jersey 08873.

Kuehn, Richard A. *Cost-Effective Telecommunications.* New York: American Management Association, 1975.

Martin, James. *Design of Man-Computer Dialogues.* Englewood Cliffs, N.J.: Prentice-Hall, Inc., 1973.

Martin, James. *Design of Real-Time Computer Systems.* Englewood Cliffs, N.J.: Prentice-Hall, Inc., 1967.

Martin, James. *Future Developments in Telecommunications.* 2d ed. Englewood Cliffs, N.J.: Prentice-Hall, Inc., 1977.

Martin, James. *Introduction to Teleprocessing.* Englewood Cliffs, N.J.: Prentice-Hall, Inc., 1972.

Martin, James. *Programming Real-Time Computer Systems.* Englewood Cliffs, N.J.: Prentice-Hall, Inc., 1965.

Martin, James. *Systems Analysis for Data Transmission.* Englewood Cliffs, N.J.: Prentice-Hall, Inc., 1972.

Martin, James. *Telecommunications and the Computer.* 2d ed. Englewood Cliffs, N.J.: Prentice-Hall, Inc., 1976.

Martin, James. *Teleprocessing Network Organization.* Englewood Cliffs, N.J.: Prentice-Hall, Inc., 1970.

Mathieson, Stuart L. and Philip M. Walker. *Computers and Telecommunications: Issues in Public Policy.* Englewood Cliffs, N.J.: Prentice-Hall, Inc., 1970.

Matick, Richard E. *Transmission Lines for Digital and Communication Networks: An Introduction to Transmission Lines, High-Frequency and High-Speed Pulse Characteristics and Applications.* New York: McGraw-Hill Book Co., 1969.

McWhinney, Edward, ed. *International Law of Communications.* Dobbs Ferry, N.Y.: Oceana Publications, 1971.

Mileaf, Harry, ed. *Electronics One-Seven.* New York: Hayden Book Co., 1967.

Murphy, Donald E. and Stephen A. Kallis, Jr. *Introduction to Data Communication.* Maynard, Mass.: Digital Equipment Corp., 1971.

Selection of Terminals and Data Protection. Cleveland, Ohio: Association for Systems Management, 1973.

Sippl, Charles J. *Data Communications Dictionary.* New York: Van Nostrand Reinhold, 1976.

Squires, T. L., ed. *Telecommunications Pocket Book.* London: Newnes-Butterworths, 1970.

Switching Systems. New York: American Telephone and Telegraph Co., 1961.

Talley, David. *Basic Carrier Telephony.* 2d rev. ed. New York: Hayden Book Co., 1966.

Transmission Systems for Communications. 4th ed. Murray Hill, N.J.: Bell Telephone Laboratories, Inc., 1970.

Vilips, Vess V. *Data Modem: Selection and Evaluation Guide.* Dedham, Mass.: Artech House, Inc., 1972.

Serial Publications

(These publications are either devoted to data communications or have a regular monthly section on data communications.)

Auerbach Data Communications Reports. Published monthly (in notebook form) by Auerbach Publishers, Inc., 121 Broad Street, Philadelphia, Pa. 19107.

Bell System Technical Reference Catalog. Published annually by American Telephone and Telegraph Co., Murray Hill, N.J. 07974 (Pub 4000). This catalog lists the Bell System publications on data communications and voice communications.

Communications News. Published monthly by Harcourt Brace Jovanovich Publications, Inc., 402 West Liberty Drive, Wheaton, Ill. 60187.

Computerworld. Published weekly by Computerworld, Inc., 797 Washington St., Newton, Mass. 02160.

Data Communications. Published bimonthly by McGraw-Hill, Inc., 1221 Avenue of the Americas, New York, N.Y. 10020.

Data Communications: A Complete Systems Guide. Published by Datapro Research Corp., One Corporate Center, Moorestown, N.J. 08057.

DataComm User. Published monthly by DataComm User, Inc., 60 Austin St., Newtonville, Mass. 02160.

Datamation. Published monthly by Technical Publishing Co., 1301 South Grove Ave., Barrington, Ill. 60010.

GTE Lenkurt Demodulator. Published bimonthly by GTE Lenkurt, 1105 Old County Road, San Carlos, Calif. 94070.

The Guide to Communication Services. A monthly updated guide to communication costs published by the Center for Communications Management, Inc., 79 North Franklin Turnpike, Ramsey, N.J. 07446.

IBM Systems Journal. Published quarterly by IBM Corp., Armonk, N.Y. 10504.

Mini-Micro Systems (formerly *Modern Data*). Published monthly by Modern Data Services, Inc., 5 Kane Industrial Drive, Hudson, Mass. 01749.

Telecommunications. Published monthly by the Telecommunications Handbook and Buyers Guide, Horizon House, 610 Washington St., Dedham, Mass. 02026.

Telephone Engineer and Management: The Telephone Industry Magazine. Published semimonthly by Harcourt Brace Jovanovich Publications, Inc., 402 West Liberty Drive, Wheaton, Ill. 60187.

Telephony: Journal of the Telephone Industry. Published weekly by the Telephony Publishing Corp., 53 West Jackson Blvd., Chicago, Ill. 60604.

Western Electric Engineer. Published quarterly by Western Electric Co., Inc., 195 Broadway, New York, N.Y. 10007.

Index